Environmental Diary:

The Obama Years

N. L. Brisson

This book is dedicated to my parents who made a huge contribution to who I am today, and especially to my Dad who gave me a love of science and who would, I think, have approved of the topic and contents of my book.

This is a view from the cheap seats

Table of Contents

Environmental Diary

Introduction

Republicans are dug in to a position which contends that humans cannot change the climate of earth. They do not even accept that high levels of CO_2 pumped into the atmosphere could cause global warming. They have generated an entire body of speculative scientific data based on core samples and temperature studies that suggests that the changes we think we are seeing are within the normal range of historical climatic changes.

However many of these studies are not accepted as good science by peers in the field. Take for example, the work of a group of people who use the title "The Right Climate Stuff". It's a clever play on the meaning of right as correct and the designation of right as an ideological position on the political spectrum. However, if you research The Right Climate Stuff on the internet you will find adherents and critics. The adherents are mostly right wing citizens, the critics are mostly actual scientists. Quoting The Right Climate Stuff data will not convince anyone who is not already a climate change "denier". The data has a partisan stink about it. Science should not depend on what political position you subscribe to. See: http://www.therightclimatestuff.com/

Science used to trump politics. And, although 97% of climate scientists say that climate change is real and is affected by the activities of people, the right wing has been able to throw plenty of shade on this bit of science by simply repeating *ad nauseum* that the data is wrong. They are also great at acting incredulous as they repeat, "We need to get over ourselves; people are too puny in the grand scheme of things to change earth's climate." In the end the right wing falls back on a philosophical argument which cannot be shaken by mere "science-y" experiments (Big Bang reference).

Perhaps looking back at the natural events, the unnatural events, the arguments, the cultural trends and the possible solutions will help. I have kept a diary of sorts of climate related news during the Obama years of 2010 -2016 (not blaming Obama for anything here). It is informative to take a look back. Perhaps you will decide, as I have, that we can set aside the decisions about whether climate is changing or not.

We could just focus on the fact that we are spinning through mostly empty space on our lovely little planet. We could remember that we have no way, so far, to get off the planet. We could accept that our planet can live on without us, but we cannot live on if our planet gets trashed or if the delicate balance of wind, water, and air that makes our planet such a delightful place to live gets out of balance for any reason.

Can't we just look at ourselves as caretakers who must keep our planet clean and happy and ticking along in the same way that we would like to keep our households ticking along. While we are arguing about what is basically climate politics the population of earth is exploding. Even if weather is not getting more extreme (which we are arguing about also) the fact that there are many more people in the path of extreme natural and unnatural events and the fact that our technology allows us to experience extreme events in real time makes such events costly in terms of both money and the psychic toll it takes on all of us.

How could it hurt to mend our flagrant polluting behaviors for a designated length of time to see if this helped matters? Well, if you are wealthy, I guess you feel it would hurt a lot, hence the pushback. If you are not so wealthy the way you feel may depend on what the media tells you, or whether or not you live in an area experiencing extreme events. However, the rather large segment of the right wing which belongs to an Evangelical religious tradition with a fundamentalist approach to the Bible should find that being earth's caretakers suits them very well given that we all once lived in that wonderful "garden" from which we were only expelled by knowledge, and that we, in that garden, apparently once lived in perfect harmony with nature.

We cannot be divided on this. If we want to know the answers we must put this to the test, be kind to earth, and see if things get less chaotic. You might excuse tornadoes that seem larger and more numerous than they once were as just being a natural cyclical occurrence, but seeing the shrinking of our ice caps, the calving off of giant chunks of ancient glaciers, is harder to be sanguine about.

Environmental Diary

Part 1

Extreme Nature

Clearly, looking back over the past six years, nature has presented us with some events of fairly Biblical proportions. Some of these events are easier to ascribe to possible changes in climate, some have more tenuous or seemingly nonexistent connections.

Scientists can measure changes in the way the wind currents move. They can chart air temperatures and compare over them with records that have been saved over time. We can take core samples of ice caps and land masses, even under the ocean. We can track insect, bird, reptiles, and animal species to chart changes in lifestyle and populations. We are good record keepers.

However, we don't seem to know what to make of the data once it is recorded. We can't decide how to use it to help the planet that is in our care. We can't even seem to decide if we should "interfere" in these matters. Actually, we often think we're helping when we are really just creating new problems to be studied and solved. I don't think we can let that stop us. Often we have produced good results.

Actually, humans love to solve problems but we don't love change. It will change our lives in many ways if we do the things we should to keep our planet from harmful pollutants that can be traced back to the ways humans live on earth. Even if we don't all agree about CO_2 emissions and global warming we have to accept the evidence that is all around us of what a mess our trash makes.

Have we altered the migration routes of birds and butterflies? Something has. Can polar bears survive without the Arctic ice cap? We'll find out. What will happen to the dispossessed walruses we learned of last spring? Will they find a new habitat? Will these changes and others affect us a lot or will we barely notice? Will new and unfamiliar critters evolve to take their places? Will we love them as we have come to love those we coexist with now?

Have we made changes to the way the jet streams aloft move, the paths they take, as a result of a warming earth atmosphere, or are we just transitioning to a new geologic era? Have changes in the jet stream pumped up our storms at times? What about the earthquakes, the tsunamis? We know that water is heavier than ice. As the ice caps melt are greater tensions being placed on the shifting plates underneath?

These Obama years (no blame accrues to President Obama) have seen horrific earthquakes, tsunamis, hurricanes, and tornadoes, even snow storms. Of course the power of these natural occurrences would not matter if the planet were unpopulated. Do they seem more horrific because of the density of humans in the 21st century and their potential for greater impacts on human life and human communities? Perhaps we can all at least agree on that.

I have asked a lot of questions because that seems to be all we have, or all we are allowed to have – questions. Politics, throughout the ages, has often registered our fear of change and forced even the most adventurous of us to adopt a reactionary position that contradicts change. People did not like learning that our solar system revolved around the sun or that our planet was not flat. Punishment is often visited on those "who dare to go where no one has gone before." After reviewing the journey we have been on (and my diary of events is hardly exhaustive) perhaps we will finally be moved to take steps that experts say might be effective to might make nature back off from recent extremes.

8.20.10

A Disturbance in the Force

(Flood in Pakistan)

*

10.5.10

Haiti Concerns

(Earthquake in Haiti)

*

3.11.11

Big Nature, Small Us

(Tsunami in Japan)

*

3.14.11

Hard to Think of Anything Else

(Japan)

*

3.30.11

Yikes!

(Fukashima Daiichi and Nukes)

*

4.20.11

Life's Ups and Downs

(Royal Wedding/ Tornadoes)

*

5.23.11

The End of the World

(Speculation about how world will end with a little help from Robert Frost)

*

8.26.11

Hello Irene

(Hurricane)

*

8.29.11

Our Hurricane Drama

(Media Coverage)

*

7.5.12

Update Emergency Preparedness

*

8.20.12

Can We Save New Orleans?

*

5.22.13

A Beast Made of Air

(Tornado, Moore Okla.)

*

11.15.13

Coastal Woes

(Philippines and storm surges)

*

2.7.14

When a Suburban Highway Became the Wilderness

(Atlanta Snow Storm)

*

10.6.15

Drowning

(Flooding in South Carolina, Fear of Drowning)

***

8.20.10

<u>A Disturbance in the Force</u>

Our attention is called to another crisis on this
embattled little planet we call home. The floods in
Pakistan are epic, apparently, and once again the
aid is slow in coming. It looks like the UN is there,
but the Pakistani government has been in disarray.
These crises have been so common lately we can
barely catch our collective breath. And these events
are hitting areas of the world that are relatively
undeveloped and inaccessible. We all experience "a
great disturbance in the force," but, after recent
experiences with all our expensive aid lolling in
warehouses in Haiti, we are leery about jumping in.
We are feeling pretty poor ourselves at the moment,
but poverty is relative and we certainly have more
than the Pakistanis. Where should we send our
checks this time?

As if human sorrow was not enough we have the
worry that the Taliban is nearby, organized and
willing to win political advantage through
geographic proximity.

10.5.10

<u>Haiti Concerns</u>

Everyone seems to believe that Americans have forgotten all about Haiti. We haven't. It is difficult to forget people who have had to live in tents for months on end and face the lack of privacy and disease and the rains.

Haiti had a lot of help just after the earthquake, although I remember how long it took to arrive because of logistics. We all felt a sense of urgency about getting food and water to the survivors. Then I believe I remember the President of Haiti asking the aid workers to step back. I remember pictures of supplies sitting in warehouses because distributing free aid supplies was supplanting the Haitian economy.

Haiti is an independent nation and apparently felt the need to take charge of the rebuilding. If we are asked to step back, I guess that means we step back.

If our government pledged money to Haiti and has not delivered it, is this, perhaps, because our aid was not welcome, or is it because we want some control over how the money will be spent, or is this just part of our own government's current dysfunction? "Curious minds want to know."

3.11.11

<u>Big Nature, Small Us</u>

I could not believe my eyes when I turned on my TV this morning and saw nature's chaos mowing down the homes, and the ordered geometric fields of Japan. A tsunami – a natural event that, since the disaster near Sri Lanka, is capable of inspiring terror in our minds and sadness in our hearts. The Ring of Fire is in motion, giant plates are moving, the earth is rearranging itself and it doesn't know or care that we have arranged ourselves on its surface to live our, apparently, insignificant lives. If this particular plate is a huge circle under the Pacific Ocean will the adjustment keep moving around the entire circle? Does the addition of melt-water from the Antarctic glaciers affect the weight on top of the Pacific Plate and cause pressures to grow deep inside the earth? These forces are so slow and the intervals between occurrences are so great that it is difficult for us to perceive if there is a time line or if the earth's crankiness is a response to our disrespectful human footprint.

Whatever the reason this area of science, this border of meteorology, astronomy, seismology, oceanography (the earth sciences) is an area where all of humanity gets to display its finer side, its more enlightened side. This is a situation where it is to mankind's advantage to share our technology without immediate concern for profit. After all, survival trumps profit.

We will go through our day with heavy hearts, as we have been doing fairly frequently lately and we will grieve with the people of Japan for the people they have lost and the order they have lost and the chaos and destruction they will have to tame once more.

3.14.11

<u>Hard to Think About Anything Else</u>

This is the second time in this decade that we have seen a modern city wiped off the face of the earth by violent natural events. It appears that an earthquake/tsunami trumps a hurricane. I could not believe what Katrina did to New Orleans, the destructive effects of the flooding and the wind were stunning. Fortunately the loss of life was amazingly low. In Sendai and northeastern Japan the damage is exponentially greater because the power of the water was enormous. Nothing could survive its force. If you picture yourself standing on that shore you cannot picture anything except oblivion. As the pictures pile up in the media you start to worry about your own safety. It was the Mayans, I believe, who predicted that the world will end in 2012. Events like this one lend more credence to this possibility. You feel such a deep sadness. Sometimes that Star Wars line about sensing a great disturbance in the force is exactly right.

On top of that we have the nuke issue. Six nuclear reactors in one area, that hits home. I live near three nuclear reactors. I never realized what could happen if there was no power; I did not realize that the core or the rods, or some part of the reactor, needs to be cooled constantly. Does it still have to be cooled after it is taken off-line? It sounds like it does. We used to demonstrate against nuclear power because we believed we did not have sufficient control over the process and the repercussions of a meltdown were serious, could not be overturned and would be long-lasting given the half-lives of these materials. It looks like we had good reasons for our concern.

So double whammy, if you weren't swept away by a ji-normous wave you might have been poisoned by radiation. These events are almost biblical in their proportions. At the very least watching the misery that nature can unleash on us should make us more determined to cut down on the misery we inflict on each other.

3.30.11

<u>Yikes!</u>

This sorry saga of the six Fukushima Daiichi nuclear reactors plays out each day in the news and pushes all our worst case scenario nuclear energy buttons. The reactor explodes (multiple times). The radiation escapes. The radiation gets into the air. The radiation gets into the soil. The radiation gets into the water. We haven't gotten to the one where the radiation poisons everything and everyone yet, but no one seems to have a grasp on what will happen, when it will stop, how to stop it, or what the long term effects will be.

It seems impossible that the heroic workers at the reactors will get out of this unscathed or even alive. What about the people who still live near the reactors? They should have been told that evacuation would be a good idea long before they were. In America, say lawsuit. How will these people be affected? Where will they go?

Once the radiation gets into the soil, water, and air it gets into the food supply. How will Japan ever recover from this? When will they recover? Will they keep the radiation from infiltrating the ocean?

I was quite certain that we did not have a clue what to do in a serious nuclear reactor failure. That I was apparently right is not giving me a great deal of personal satisfaction. We will all watch this play out with great trepidation because this chain of events could happen here. I wish we could shut down all the world's nuclear reactors until we figure this out. It is potentially too toxic, and there are no true remedies. I also know we are unlikely to abandon nuclear power right now when we need to lessen and end our dependence on fossil fuels. So, yikes!

4.29.11

Life's Ups and Downs

This morning I feel like I am watching one of those movies that juxtaposes the best and the worst parts of our lives and that has our feelings moving up and down the emotional continuum as if we were on a serotonin roller coaster ride.

On the one hand we were treated to a very lovely wedding that managed to be royal and to maintain a certain sweetness at the same time. Kate and William had such down-to-earth demeanors; their smiles were so genuine; they managed to look like two young, but quite mature people setting off on a great partnership. Her dress was just right I thought, smooth and stylish, satiny and trained, with a lacy neckline and sleeves, the simple veil with the small tiara, perfect. No meringues were in sight. Commentators were shocked that she and William kissed on the lips, but the kiss was also so like the kiss of any loving couple at their wedding that it added to the graceful and gracious quality of the whole event.

On the other hand we wake up to 300 people dead in southern America from tornadoes. CNN bounces back and forth from covering the wedding to talking to tearful people who have lost everything. They are covering the President's tour of the terrible destruction. Parts of Alabama, Georgia, Tennessee, and Mississippi look like war zones. People are losing some of their resilience. They are not sure where they will go or what they will do. In moments their life, as they knew it, is gone. All their possessions will not be available when they start over. They will have to start from scratch. Our hearts understand their sadness, and we look to our agencies and our government to start them on the road to their new lives. We hope if or when this happens to us we will also have these same resources to depend on.

Back we go to Kate and William leaving Westminster Abbey in their little convertible car and the streamers flying off the back. Perhaps we worry about whether true love can survive in a fishbowl. Will these two learn how to fly under the radar of the media and the paparazzi? I hope we don't hear about every move they make every day of their lives. I wish them every happiness.

We also have the second to the last space flight beginning, with the shuttle on the launch pad. We have the fears that some other nation will get to space before us because we can no longer afford a space program. We have the poignant pleasure that Gabby Giffords will be recovered enough to go and watch her husband leave for space. We have Elton John's "Rocket Man", a perfect soundtrack for the occasion.

Our lives on this day are torn between beauty and sorrow, between hope and fear, between optimism and suspense.

5.23.11

<u>The End of the World</u>

Robert Frost once wrote about the end of the world in *Fire and Ice*. You remember this poem, right?

Some say the world will end in fire,
Some say in ice.
From what I've tasted of desire
I hold with those who favor fire.
But if it had to perish twice
I think I know enough of hate
To say that for destruction ice
Is also great
And would suffice.

This poem comes to mind especially after we had the weekend of the "Rapture" that wasn't. I don't want to be too flippant about this because I do not want to tempt fate. Frost suggests that our emotional vices make either fire or ice a fitting end for us very flawed humans.

However, after watching weeks and weeks of damaging tornadoes and the new damage to Joplin, Missouri, I wonder if the world could end in wind, and just what emotional dysfunction Robert Frost would connect to apocalypse by wind.

Perhaps the wind comes from the current climate of dissent. I know it could be global warming but that doesn't fit the poem's paradigm. It is not a human emotion, although it may result from a human need for comfort and luxury (which are considered sinful). I do not think the people of Joplin are especially guilty of any of these human failings and we are all in awe of how fearful nature can be. We've seen almost all of nature's scary power the past few years. What can we say to people when they are faced with these kinds of loss? You are with us in our hearts and in our minds. It isn't enough, but it is all we have.

8.26.11

<u>Hello Irene</u>

We are waiting to see where Hurricane Irene will go. This is a key difference between a tornado and a hurricane. A tornado seems to pop up out of nowhere. It gives those in its path little time to prepare. When it comes in the middle of the night an emergency alarm may jolt you awake if your community has one. If it comes by day the quality of the light in the sky may tip off people who can actually see the sky and they may pass on the warning.

With a hurricane we have plenty of time to watch it wobble its way toward us or towards someone else. Irene seems a little slower than some. We can see her coming but we still have uncertainty on several points such as what her exact path will be (which we really wish we knew), exactly how strong she will be, exactly when she will arrive, and how much damage she will do.

I don't know if there is much choice between being hit quickly by the knock-out punch of a tornado or the drawn out anticipation of a slow-moving hurricane. Once the damage is done they probably feel pretty much the same to those who are hit; so it's either the drama of lengthy anticipation and dread of the hurricane, or the lightning-fast zap and heart -attack funnel of a tornado.

I am usually soothed by the fact that I live in a part of the country where tornadoes and hurricanes rarely affect us. I was eight when Hurricane Hazel hit and it was a powerful enough storm that I still remember it. I remember the power of the wind and the rain and I remember the calm of the eye. I remember having to cook out in the back yard on the old cook stove that was there at that time and I remember how difficult it was to find staples like candles, bread and milk. People didn't buy water then.

I have never experienced a tornado although one Labor Day in the recent past we had a very dramatic hit from a storm with a sky that was sheeted with lightning for well over an hour and lots of straight-line wind damage. Trees were down everywhere but only on east-west streets. North-south streets had no damage.

Given these two experiences I would not chose either type of storm and, I assume, neither would anyone else. I don't think this storm will be another Hurricane Hazel, but there is at least one potential path that takes it inland and could put me in the storm.

Maybe I should buy a few batteries and some water.
NOAA says this is not just a coastal storm, that its
impact will be felt well inland. It could still go out
to sea (although the prospects for that look very
iffy). They are making this storm sound pretty
scary. Good luck everyone.

8.29.11

Our Hurricane Drama

Irene was a stunning experience. Not the strongest hurricane in nature's arsenal of significant weather events, but she certainly demanded the attention of at least 2/3 of the East Coast of America and of a portion of coastal Canada also. It is our great good fortune that this hurricane, on this track, was not any stronger than it was. Something about the human spirit needs excess. If a hurricane is Category 4 we secretly feel a certain grim satisfaction. If we prepare for disaster and it doesn't come, we know we are wrong, but we feel vaguely let down. This is a tiny portion of our human nature and a sort of bizarre aspect we are not really proud of. Our more rational self says this storm was quite large enough, thank you.

I say this storm was large enough to justify the preparations taken by our state and local governments. I already hear people complaining that the lead in to this weather event was a perfect example of overkill and given that every channel on TV had 24 hour a day coverage it would seem to be true. Actually this is more a function of channels dedicated to all news all the time and we must admit we were fascinated. There wasn't a lot of other news going on with our government on hiatus and, although it was a bit excessive, we couldn't look away.

I'm sure that every part of us, except that little portion that is titillated by catastrophe, is happy that a hurricane like this which hugged the East Coast for so many hundreds of miles was not as strong as we feared it would be. If I was one of those who was evacuated from my home and then felt that I had been inconvenienced for nothing I might feel differently. I was, however, happy to see our dysfunctional government is not so bad at the state and local level and was able to work in concert when necessary. That was reassuring.

Now we will have to spend millions of dollars that we don't have to clean up from this hurricane/tropical storm, but not as much as we would have had to spend if it had been more powerful. Now we have lost millions in tourist dollars that would have contributed to the economies of the affected states, but tourist services will be back on line much sooner than they would have been if the storm had lived up to its early hype. Let's count our blessings, thank our leaders and our newscasters and move on to cleaning up and living our lives.

7.5.12

<u>Update Emergency Preparedness</u>

I must say it was a bit upsetting to watch the DC area including Virginia, West Virginia and Maryland trying to recover from the surprise storms that hit them last week. A huge number of people lost power at a time when there was an on-going heat wave. The power company had to slog through neighborhood after neighborhood clearing debris and restoring power and although they obviously worked hard and worked long hours their progress did not meet the needs of the residents of the area or give the rest of us any kind of confidence that progress would be any better if this happened near any other large metropolitan area.

Yesterday, as we all watched parades and picnicked and celebrated, news channels reported that some areas without power were also without food. I don't know what happened to the non-perishable food supplies, but lack of refrigeration is what happened to the perishable food supplies. Stores and individuals had to throw away their frozen foods once their freezers shut down and they certainly could not schedule any deliveries until their power was restored. Some people, especially in West Virginia, are living in very remote areas where power may not be restored until tomorrow. Today's news reported that 90% of people now have power, which means 10% do not. People had trouble getting water, cleaning themselves, cleaning clothing.

Our weather is showing a tendency to get more violent, rather than less. We are not set up to live without power at all. Most of us don't own generators. Most of us don't have a backup well or stream or an emergency outhouse tucked around the corner. If storms hit metro areas housing millions of people, perhaps amenities need to be rolled out by some agency like FEMA so that life can go on until power is restored. Bring in generators and freezers to neighborhoods where people can store food, bring in port-a-potties, bring in portable showers and water trucks. We may need emergency supply kits that include big items like this which can be mobilized as needed.

I think it is safe to assume that if this happened once it will happen again, and, in fact, it has not just happened once; it has happened over and over again recently (especially those tornadoes). We need to reassess our emergency preparedness and we need to update our offerings.

8.30.12

<u>Can We Save New Orleans?</u>

Can we save New Orleans? Can we afford to keep saving New Orleans? It seems clear that it is "iffy" to build a major city on the shifting ground of a river delta. We love New Orleans. Even if we haven't visited we love the idea and the vibe of this jazzy city. There will probably always be some kind of city in the Mississippi delta because of shipping concerns. Clearly some of New Orleans is on high ground and can survive any number of windy/rainy disasters. But the parishes that occupy low-lying land may be too expensive to maintain, especially if our weather keeps getting more severe.

 The Army Corps of Engineers is getting a lot of praise for how the new levees and pumps are functioning. But Isaac is only a category 1 storm. It is a fairly stationary storm however, and is lingering over New Orleans and dropping *beaucoup* inches of rain. What if this was a category 4 or a category 5 hurricane? If we had followed the Dutch designs and built a truly strong system we would not be facing this continuing uncertainty, which we could face again in any hurricane season. The Dutch design with its hydraulic doors was extremely expensive and I assume that is why we didn't decide to go that way. In the long run the Dutch design may be the only thing that could preserve the city of New Orleans in its current form.

Meanwhile, we are sorry for the suffering and fear that the people of New Orleans are experiencing on the 7th anniversary of the terribly destructive hurricane, Katrina. I hope when the rain, wind, and tides subside we will find the city and its people weathered the storm well. We already know that at least one parish was inundated, that people feared for their lives and had to be rescued from their attics and roofs, but this was not a parish that benefited from the Army Corps of Engineers upgrades.

5.22.13

<u>A Beast Made of Air</u>

I still cannot get used to having natural disasters happen right in front of my eyes in real time on my TV. On Monday, May 20th, we watched a tornado approach an intact town, Moore, Oklahoma. We watched as the storm chasers in the helicopter focused on storm clouds. We watched a powerful tornado whirl out of those clouds and tear across the earth towards that little town, Moore, Oklahoma. That throbbing up-elevator of air was inescapable, totally unstoppable, and it was impossible to predict what its exact path would be. A tornado is a fearsome thing, made from wind and air and pressure and that helicopter showed us the true nature of the beast as it ripped and snarled its way through the sturdy modern homes made by humans who understand the beast, even if we cannot control it.

Even as we watched, the people in this small town were told to seek shelter. They were told that an interior room was not secure enough. They were told to find a storm shelter built under the earth. As they huddled wherever they felt somewhat safe, we watched the tornado pass over Moore. We could not see the town or the huddled people but we tasted the fear they must be feeling; we tried to picture where we would seek refuge if that was our town, our house, our school, or our mall. We could see the image of the muscular tornado coming to us from the cameras on the helicopter. We were informed about the debris field. We held our breath. It didn't take long. We saw the tornado turn into a narrow rope that climbed from the earth to the sky and we were told that the tornado had "roped out". It was gone almost as quickly as it had formed.

And then we followed the cameras into Moore, Oklahoma and we could not believe what the cameras captured. A huge swath of Moore was gone. A long narrow path of buildings throughout the town had been taken apart like a set of Tinkertoys. The cameras almost immediately located the elementary school and we learned, from an understandably emotional local newsperson, that the school had been destroyed. That meant that children might be injured or dead. We waited with the residents for the death toll (24), for the stories that people who lived through the tornado had to tell, and we knew that we could not experience the way the post-tornado air felt, or the anguish of parents as they waited to be reunited with their children, or the realization that there was no home to go home to.

In the aftermath of the storm, when all that is left is picking up the pieces, we become bystanders again. We become bystanders who wish we could reach through our TV's and put everything back in its rightful place and bring all those who lost their lives back to the arms of their loved ones. All we can do now is pray and contribute to the Red Cross, but at least this is something. Despite all of the progress mankind has made, extreme weather leaves us almost as helpless as primitive man.

11.15.13

<u>Coastal Woes</u>

We are becoming inured to coastal inundations and it looks like these awful disasters, that take thousands of lives, flatten towns and villages, and cost countries millions and billions of dollars in aid to affected citizens for relocation and for rebuilding, will become more frequent. Here we have the case of the Philippines, hit so hard by a typhoon that the entire infrastructure which would normally allow relief to inhabitants has been devastated to the extent that even basics like water and food are almost impossible to provide. There are no warehouses for safe storage. Residents, pushed by the desire to survive this natural disaster, are looting supplies, hoarding what they find and making it more difficult to provide in an equitable manner to all who were affected and survived. This must be very similar to what happened in Sri Lanka and almost like what happened in Japan. In these places the population had to leave the affected areas and this also was what occurred in New Orleans after Katrina and in New Jersey and Long Island after Sandy.

I am thinking that we almost have to accept that the level of water in our oceans is rising and we have to prepare for how this will affect our coastal populations and cities. I am speaking globally here. Losing so many people in each of these disasters leads to profound levels of global grief. We begin to think about apocalypse. We are a smart species, with big brains. We should be able to see the evidence in front of us and start to move people back from vulnerable coastal areas.

If I lived near the ocean I wouldn't want to leave my property and face the possibility, that without legal enforcement, others would eventually take my treasured spot. Governments, however, cannot afford to keep bailing out people who insist on living in harm's way. Aren't there ways to protect someone's property ownership even though the property is considered uninhabitable at the moment? Will we compensate people when their land disappears under water or is above sea level only intermittently?

I don't think coastal residents can expect such recompense if the cause is climate change, something beyond human control at this point. Governments probably cannot afford to compensate everyone who owns beachside property for dwellings they have built there. Perhaps governments might be better served if they paid to relocate coastal residents and left beaches uninhabited. (OK, horribly unpopular – I can hear the outcry already.) Japan may not be able to rebuild the villages hit by the recent tsunami because of nuclear contamination for decades. Sri Lanka has recovered somewhat, but their build up was far less dense and homes there are less structurally complicated because they are more suited to the climate. New Orleans is still not back to normal after Katrina and towns affected by

Sandy are nowhere near full recovery. If another giant storm hits any of these places any time soon rebuilding will probably become impossible.

So far the problems on our seacoasts seem to be when there are surges following huge storms. But we have to ask 1) are the storms we are seeing supersized, and 2) are the surges reaching farther inland than previous surges? I was reading about Norfolk, Virginia, an important port for military ships on the east coast. In Norfolk the water levels of the ocean and bay have risen several inches, but predictions say that the water level will probably rise by at least one meter (39 inches), which will change the entire configuration of the port.

Climate deniers in Virginia do not believe these predictions and refuse to devise contingency plans. Here's a case of a section of our coastline we can keep an eye on to see what actually happens. However, if we are concentrating on one small section of coastline, disaster will surely strike elsewhere. What if a major population center is involved as with New Orleans? Shouldn't we be building Netherlands-style protections around our major coastal cities? Shouldn't we begin now to take some action along our less populated beaches to create unoccupied flood zones so we will not have to lose any more people to unruly seas?

http://wwwp.dailyclimate.org/tdc-newsroom/2013/09/virginia-rising-seas

http://articles.washingtonpost.com/2012-06-17/national/35459771_1_sea-level-rise-sea-levels-hampton-roads

http://www.wavy.com/news/local/norfolk/odu-researches-sea-level-and-superstorms

2.7.14

<u>When A Suburban Highway Became the Wilderness</u>

What will we do about that storm-induced traffic jam in Atlanta, the one that trapped people and cars overnight on multilane highways and probably cost millions of dollars to untangle, perhaps billions? I suppose at this point that is up to the people who govern Atlanta and Georgia. But that was only 2 and ½ inches of snow and because everyone tried to get home at the same time chaos ensued. If this is a once-in-twenty-year event, nothing will happen.

However, given the wacky weather we are experiencing these days (for whatever reason) these scenes (at the least uncomfortable, at the most deadly) may play themselves out over and over in future winters and other cities. We can't know for a while if this will become a "thing". We won't know for a while if this is an indictment of our aging infrastructure.

We would hope that Atlanta and the state of Georgia might come up with a plan to do staggered releases before a storm, or to do shelter in place warnings while the roads are being cleared. Once everyone leaves wherever s/he is and gets in his/her/their vehicle to go somewhere else it is too late. The traffic on those highways was so great that the plows were blocked and the salt could not be spread and gridlock resulted. This is a new travel nightmare that I hadn't even thought of yet, but which I will now add to my repertoire.

I was visiting my sister for a few weeks in Greensboro, North Carolina because she was getting ready to move back home. There was a snow storm which to me seemed like a dusting. The snow melted in the day and then the temps took a dive and that little bit of snow turned into sheets of ice. As long as the temperature stayed below freezing, which in this case was about a week, nothing moved in Greensboro. Even walking was hazardous.

My sister and I could not get out of the apartment to buy the boxes she needed for packing and we just had to cool our jets and pretend we were snowed in at a lovely ski chalet. Actually it forced us to have a nice relaxing visit and the heat did not go out or the lights so we were fine, but it did put a crimp in our schedule and it did force us to decelerate. We finally got the boxes and we were able to do some packing before I had to leave but at the time I remember being shocked that such a small weather event could have such a disproportional effect. However, what happened in Atlanta was exponentially more problematic.

This week we got hit by the western edge of a storm that came across the middle of America and up the coast. Areas nearest to the Atlantic got the worst of this snow and my nephew and his wife and family in Pennsylvania are still without power four days after the storm. Obviously there are people who were hit harder than my little northern city was by this storm. We got about twelve inches of snow and driving was treacherous all day, but our streets are generally cleared and salted as soon as the snow stops and this storm stopped midday, leaving rush hour less fraught. People were also released from work on a staggered schedule. We have lots of practice with snow. We don't get a lot of ice. Ice is a killer. There are few vehicles that do well on ice. So I don't think we get any bragging rights about not having storm-gridlocked highways, but Southern states should feel free to pump our Highway Department officials for any winter road wisdom they can adapt to their situations.

10.6.15

Drowning

We are experiencing weather so extreme that we are unprepared for it. There are protocols for hurricanes and tornadoes. But when we experience something that is called a 1000 year rainfall, and when we haven't even been keeping records that long, there are no protocols.

Here is a nightmare scenario. Families and individuals are battened down at home, at work, or even in bed for the night and the living room, the office, the bedrooms start filling up with water – fast. Almost before you realize it the water is to your knees, mid-thigh, to your waist. You have to get out, get everyone out, leave the house – but water is just as high outside the house. Your car, now useless, sits in the driveway or in the parking lot. It is still raining steadily with no signs of stopping.

Somehow you do get to higher ground, or someone with a boat comes and helps you out, or the water, which flooded in so quickly, flows back out to flood a downstream neighborhood. Perhaps you were already in your car. Perhaps it stalled in the deep water and you have to get out and find a way to survive. I cannot imagine how people keep their heads in such circumstances, but they do, and they make it through, although this time at least nineteen people died. We recognize this feeling from Katrina.

Drowning seems a hard way to die because perhaps you see and feel the water overtake you.

Officials did not have a grip on what would happen given such a quantity of rain falling so continuously. They didn't predict the failure of all the earthen dams on the lakes because the amount of rain was unprecedented. This part of South Carolina is called the lowlands after all. They did not order an evacuation which they might have if they understood the magnitude of the downfall.

It appears to me that the number of nor'easters has been increasing. The whole East coast has been subject to more flooding, more erosion, more precipitation. I have no proof that this is so; I don't have the data. There is proof that the water is rising in the Norfolk, VA area and that land is being lost as the shoreline pushes inland. I used to long to live near a beach, but I am starting to appreciate the safety I get living away from a coastal area.

As we watch your plight on the news we are thinking of you, the people of South Carolina, whose lives have been turned upside down by water, usually such a benign and live-giving substance, but, as we know, sometimes also such a destructive force. We are sorry for all that you have suffered and all that you have lost and we are happy for those who made it through this deluge.

Environmental Diary
Part 2
Old Energy, Pollution, and Water

Pollution did not begin in 2010, and old energy (fossil fuels) did not gain recognition as being too dangerous to keep relying on originate in that year either. We have been unkind to our oceans, streams, rivers, and lakes since the Industrial Age began. 2010 is just when I started keeping my somewhat sporadic diary on the subject of the environment. These problems were not suddenly solved in 2015 as you know, but the 2016 election sort of eclipsed a lot of news and grabbed my attention. After the election we will have all the same environmental concerns to fight about, issues that will not be resolved in any satisfactory manner unless both political parties get on board.

We know many reasons why we cannot work on cleaning up our act in a unified crusade anytime soon, at least not without divine intervention, which could very well come in the form of some cataclysmic event that literally rocks our world.

First, fossil fuels have been considered essential resources for centuries but their importance increased exponentially with the advent of the Industrial Age, central heating, the automation of factories that arose from the innovative idea of an assembly line often attributed to Henry Ford. Factory owners, these days, heads of corporations, do not want the Industrial Age to end. It has been their "cash cow" for decades. And it is true that we will probably always have to manufacture things.

Nations and individuals have built their lives around locating and tapping sources of fossil fuels. Underneath us all lies that deep (or thin) band of fossil fuel materials laid down by dinosaurs during their own cataclysmic demise.

The Middle Eastern nations do not manufacture much but they have gotten used to owning most of the world's biggest oil deposits. This has given them great wealth and plenty of clout among nations that are far larger and more metropolitan than they are, given their location in desert lands (although they were once home to thriving cultures and they are still the birthplace of at least three of the world's great religions). So there has been a certain balance of power among the planet's nations that we have become accustomed to.

That balance is already changing as other nations are finding and exploiting their own oil reserves. However these oil and gas reserves are difficult to access, thus we must suck them out of century's old shale layers laid down beneath our land, a process with a number of (arguably) bad did effects.

- Water that lights on fire
- Earthquakes, as in Oklahoma
- Spills
- Long supply lines (pipelines)
- Oil trains that can explode
- Human activity on pristine land that often affects an entire ecosystem

Industrialists are happy about this new oil rush because it tips the balance of power back to Western nations. This shift would be more enjoyable if Western nations were still centers of manufacturing. Alas, we are not. Factories, at least old style factories, have moved south or east. All those newly found oil reserves have to be refined and, for the most part, shipped out to the place where the Industrial Age has gone on its world tour.

Still money lines Western pockets and bank accounts from fossil fuels and those who profit don't want the fuel gravy trains to stop. How much of the pushback against climate change, global warming, and exploring alternative energy sources comes down to money and how much to fear of change and how much to balance of power – hard to say, but that gives the wealthy lots of reasons to root for the status quo. It may even mean that profit trumps truth, but that should not surprise anyone.

The truth is that we do not have enough alternative sources of energy to totally replace fossil fuels as things stand right now. We do not even have great technology to turn pollutants and trash into byproducts that are either harmless or useful when recycled. And once we are in for a little, it ends up being that we are in for a lot and back to using fossil fuels at levels that are normal or even greater than normal.

This part of my diary begins with the BP oil spill, a nightmare that sent deadly oil into the pretty much pristine waters of the Gulf of Mexico and which went on and on because no one had invented a part to plug the leak in the main oil pipe under a deep water drilling operation. We know all about how to clean wild life covered in gooey, oily gunk, although we don't always hear the death toll. We still have dish liquid in our supermarkets that advertises its usefulness in cleaning up "poisoned" ducks.

Fracking pros and cons and the Keystone Pipeline have earned a place in this diary. My main objections to fracking are that we cannot afford to put oil or gas into any of our precious fresh water sources because we cannot get it back out. We do not have cost effective technology to make water polluted in this way potable once again.

And then there is the matter of our oceans which is scary enough to keep you up at night. Our oceans are part of earth's water cycle. No one out in space is pouring new water down on earth. We cannot make water without depleting our air. The water we have is a finite quantity and keeping it circulating is what allows for life on earth (or any planet). If it eventually circulates water that is so tainted by pollutants that the very rains and snows that fall from the sky become "killers" (think acid rain) we will have to leave earth or die. How important will money be then if it can't buy life?

Whether you believe that we are in dire straits or not, shouldn't we use some of our highly valued human ingenuity to keep our planet clean and livable? It seems like a no-brainer to me, but then I am sitting in the cheap seats.

5.2.10

Eco Terrorism? (BP)

*

5.3.10

Oil Spill 2 (BP)

*

5.11.10

Oil Spill 3 (BP)

*

5.20.10

Oil Spill Variation (BP)

*

7.16.10

Oil Spill Reprise (BP)

*

4.19.11

Scientists vs. Locals – The Health of the Gulf

*

5.16.11

Hydrofracking

*

6.28.11

Goodbye Whales, Goodbye Beaches

*

10.19.11

Oh Snap! (Hyrdrofracking)

*

11.7.11

Oil and Water Don't Mix

*

1.12.12

Dependence on Foreign Oil

*

4.16.12

Waterworld

*

12.10.12

Fracking for Dollars

*

12.11.12

The Keystone

*

1.31.12 or 2.3.13

How Likely Are Water Wars?

*

10.9.13

Stonewalling While Our Oceans Die

*

4.24.14

Don't Bogart that Planet

*

8.20.14

Sonic Cannons Kill: Save the Oceans

*

12.10.14

Gray Mountain by John Grisham – Book

(Mountain top coal mining)

*

2.1.15

Vote No on Keystone to Save America's Fresh
Water

*

5.12.15

Choo Choo Kaboom

*

5.21.15

Sea Sick

5.2.10

<u>Eco Terrorism?</u>

I am no longer a trusting soul, although I used to be. I always seem to imagine the worst. I am a suspicious and skeptical person these days. I see two environmental nightmares occurring in very close proximity. I recall that we elected a president who said he wanted to lessen our dependence on foreign oil. Most Americans seemed to agree with this. When Obama said he wanted to use a mix of different approaches to beef up domestic energy sources he started to lose voters here and there. He mentioned nuclear. Some voters are staunchly against increasing the use of nuclear energy. Clean coal - some voters believe there is no such thing. Increase offshore drilling - another group of supporters makes it clear they do not back this strategy. At least Obama never seemed in favor of drilling in ANWR. We do have a dilemma that speaks to what our economy will be like in the future, how much we will be able to produce in the future, and how we will keep our preeminent place in world politics if our economy doesn't stack up.

I can't help wondering if, perhaps, these terrible "energy-related" accidents could be the work of some eco-terrorist group who wants to block Obama's homegrown energy program because they think he's wrong. Instead of telling him he's wrong, maybe they decided he needed a demonstration. If this isn't people trying to make their point, then maybe the cosmos is having its say. The American energy problem still remains, unsolved and very real. We may still have to use some of these very dangerous technologies unless we can come up with some kind of global energy clearinghouse that shares and rations the world's resources and thus, effectively, takes energy out of the power equation. I wish we could do that before we have nothing left that can fly, swim, or sing.

Can you be cynical and idealistic at the same time? Apparently. But, I guess, at the very least these thoughts officially add me to the ranks of the "nut burgers."

5.3.10

<u>Oil Spill 2</u>

This oil spill is not like Katrina. Katrina was an unprecedented natural event compounded by human error in which a really large city was destroyed by a really large storm. There were people involved, lots of American people and they were stranded without the attention they should have had. The amount of time it took to visit and rescue those people was unconscionable. This was a problem you could throw money and resources at. But our government couldn't decide whose resources should be tapped. Those people almost died because of what was essentially a turf war. Rescue first, turf war later is the lesson we should have learned from this.

The oil spill is different. I watched on the Internet for 4 or 5 days while news sources said there was no leak. Even I didn't believe that and I know nothing about undersea drilling. If the industry had not been in denial they might have been able to tackle the problem before it was so large. It's even scarier to think that oil companies don't have a fix ready for when this kind of thing occurs.

Now we are going to see those sad, sad birds covered with slick oil and we'll see beautiful beaches soiled by oil and healthy wet lands suffocated by oil. What we won't see is the people who lose their jobs, although I'm sure we'll hear the numbers. Seafood could be quite scarce for a while. These are not houses built by humans that can be rebuilt by humans. This is our beautiful Earth spinning in the emptiness of space, our lovely world that is our only home damaged once again by us.

Nature is resilient - it may be able to repair itself, but it will take a long time and we may reach a "tipping point" where nature no longer bounces back.

This is a turf war between government and big business. The Big business in this case is supposed to be an expert in oil technology. Obama can't fix this. He has to rely on experts. He can throw money at it and people to help with cleanup, but cleanup will not be effective until this oil stops flowing. The well has to be capped. Obama can't cap the well. He will have to rely on either the BP Corporation or maybe the Army Corps of Engineers.

No this is not like Katrina. There are no stranded people. But it does hit a fragile area that may be even more damaged than last time.

5.11.10

<u>Oil Spill 3</u>

There are so many things that I can watch but cannot change. I am experiencing such grief over this oil spill. I watch the oil moving towards the gorgeous marsh lands and the barrier islands and I want to be omnipotent. I want arms that will reach out and hold the oil away, I want a giant vacuum cleaner that will suck the oil off the water. I was so hopeful that at least one of the bizarre ideas suggested by BP Petroleum would work and this disaster would not happen, just because it's too awful to believe that it will happen. It's pretty clear now that there will not be a miracle solution. We can see for sure that there are people who are willing to put our planet at risk without any knowledge of how they will mitigate possible damage.

I suppose everyone can argue that we need the oil and that it was all done, not for greed, but out of patriotism, to end our dependence on foreign oil. After all I am guilty too, the queen of central heating. I have not offered to give up my personal comfort. However, apparently we are playing with technologies we are not in control of. Will a nuclear accident be next? How many birds will die? How long will it take for the beaches and marshes to be free of oil? How long will it take before the oil enters the gulf stream and travels out into the Atlantic and even over to Europe?

How long before the sea recovers and the fish and the shell fish? Will it ever recover? I thought the planet would be ruined at some distant time in the future. We have all been watching the destruction in slo-mo, but it wasn't like this. All day whatever I do in the back of my mind I am following the progress of the oil, wishing this would be the moment the oil stopped flowing into the Gulf of Mexico, and each day when the day is through the potential for carnage is worse.

5.26.10

<u>Oil Spill Variation</u>

We haven't offered to help BP or the people in the shellfish industry or the people who have to clean up the animals and the beaches. We haven't held any concerts, no celebrities have been visible, organizing fund-raising efforts.

Most likely there are several reasons for this:

1. We don't know how to clean up the ocean or the animals and feel we should leave this up to the experts.

2. We don't like our corporations. We see them as rapacious and as unreliable partners, and they have been both. We may even look forward to seeing them losing millions/billions.

3. We have had so many crises to deal with lately that our pockets are a little bare.

If we get behind BP and help them with contributions the way we do in other catastrophes maybe the BP people would feel a partnership with us all, instead of just having to swallow the blame (and, I don't know about you, but I do blame BP). Instead of having to act defensive and prideful maybe they could put all their energy and resources into the task at hand – getting the oil shut off and cleaned up. Perhaps it could be a matching funds arrangement or some contract that still required that BP use its own money in addition to ours.

When they were providing a product we needed we liked them well enough. We suspect they have been greedy since they posted huge profits while our gas pumps prices continued to rise. They have been heedlessly careless of our environment with their "need for speed." It might be a good thing to have them "owe" us.

It does us no good to withhold our charity from BP because they will refuse to feel the shame we wish them to feel. They are much more likely to come up with creative and effective solutions if we are behind them. Maybe we can at least try to encourage a new, more responsible corporate style. Maybe allowing another corporation to fail or sending another corporation running for the hills is not in our best interests. Besides, while we are busy blaming BP a whole important ecosystem is dying.

7.16.10

<u>Oil Spill, Reprise</u>

It was surprising to find myself spending a day (once again) wondering what was going on with the spilling oil. I did not expect any action until August except for the daily cleanup of beaches and sad animals. Suddenly they were going to fit a new oil cap and test it for pressure to see if the oil line is whole or if it is leaking elsewhere. But, as is often the case with constant news coverage, on Wednesday it was "much ado about nothing". For safety reasons the pressure tests were postponed for 24 hours. Now they decide to be cautious! It's OK, as long as they will remember to choose safety over cost cuts going forward. We'll see. Our long term memory is really bad.

So it's Thursday, the cap has been fitted and the oil has stopped leaking into the Gulf. There is no oil leaking into the Gulf. Even though we know it is a temporary cessation we will take one minute to celebrate that beautiful absence. Hooray! The oil has only been shut off for about two minutes and someone with a Southern accent is asking if we can now end the moratorium on drilling.

Stage Directions: Smack your forehead with your open palm and say, "duh."

I know these guys are really talking about jobs. I know we will go back to drilling, although perhaps not in such deep water. We have not replaced oil with other energies yet. Let's at least wait to start drilling until we get this oil spill stopped for good and until the Gulf is well on its way to being clean. Let's wait for the day when that ugly cloud of oil stops pouring from that well head for good.

4.19.11

Scientists vs. Locals - The Health of the Gulf

Scientists say, according to a recent news story, that the health of the Gulf of Mexico is nearly at pre-spill levels. Although the Gulf did get the heroic efforts of thousands, equal to the labors of Hercules in the Augean Stables, it does speak to how lucky we might be if our Earth is so resilient. I was thinking we had done something so catastrophic that our planet would suffer permanent damage and yet somehow we may be able to limp back. Perhaps the waters will not be quite as pristine as before the BP spill, perhaps the critters that live in the sea will be toxic for a while or will suffer extinction or mutation. Perhaps the margins of the gulf, the shallows where the birds live will exhibit signs that all is not well at some future time. CNN reported that the Gulf bounce-back earned a score of 68 out of 100. In high school this was barely passing.

On the evening news reporters were back in Louisiana because it is one year since the spill first occurred. The residents of Louisiana were not nearly as positive about conditions as the scientists were in the morning news. Residents interviewed said that oil still washes up on all the barrier islands; they showed what happens if you dig down into the sand where just beneath the surface bands of oil are clearly visible. The toll on local families has been great and the fishing businesses are nowhere near back to normal.

This was an oil spill of 200 million gallons of oil over a period of 85 days. It seems impossible to believe we could come back from this event this quickly. I am holding on to some disbelief, especially after listening to local people, but I am willing, given better evidence than I have currently seen, to be convinced that a full recovery is possible and will happen sooner rather than later. One does wonder how many more events of this magnitude our planet can endure.

5.6.11

<u>Hydrofracking</u>

We live near a geological formation called the Marcellus Shale. Where there is shale there is some oil. Oil puts dollar signs in people's eyes. It put some in the eyes of local residents and it put some in the eyes of the oil business. The method that is used to remove the oil is controversial. Water and chemicals are pumped down between layers of shale so that the oil or gas can be ferreted out and lifted to the surface.

The residents, in spite of the dollar signs in their eyes, still had the sense to do some listening. They learned that in Pennsylvania chemicals from hydrofracking (as the process is named) have leached into the ground water making people unwilling to drink the polluted water, even though the company extracting the oil says the water is safe.

The farmers and landowners in the Marcellus Shale who had dollar signs in their eyes started getting cold feet. They said "no" to the oil people. The problem is that the oil people don't want to take "no" for an answer. They are always presenting experts at local meetings who want to reassure the landowners that hydrofracking is perfectly safe and will not harm their water. The farmers know better than to believe these people. They still keep saying "no."

But some have always believed the oil men and they are still seduced by the dollar signs. This gives the oil guys some hope. So they don't give up. They push, they keep advertising meetings, publishing articles, making commercials. They bring the issue up over and over, each time hoping they will get a different answer. When is an issue done? When does "no" mean "no?"

Probably they will give up only if new sources of power overtake fossil fuels. It is a hard thing to watch though. We get our water from a pristine lake very near the Marcellus Shale. What the dollar sign people decide may have an effect on many nearby residents. Those of us who own no land in the shale may inherit polluted water but will not benefit from any profits and no amount of profits would make up for a compromised water supply.

6.28.11

<u>Goodbye Whales, Goodbye Beaches</u>

Apparently our oceans are in bad shape and are on the verge of "mass extinctions – of fish, whales, and other marine species. An article in the on-line publication *The Week* says "our oceans are under siege from a deadly trio of threats – rising water temperature, acidification and lack of oxygen – that played a role in similar mass extinctions 55 million years ago,"(according to Jelle Bijma of the Alfred Wegener Institute). The culprits are overfishing, pollution and run-off of fertilizers, the article goes on to say, and also the fact that oceans are absorbing most of the CO_2 we pump into the atmosphere which causes acidification.

We would have to stop overfishing, reduce run-off and eliminate CO_2 emissions. We are highly unlikely to do any one of these things. Good bye kelp forests, goodbye coral reefs; what will our beautiful oceans be like?

Can we come up with a filter or something to take bad things out of the ocean? Can we nurse our oceans back to health like we do when we devastate some environment on solid land? All the oceans are interconnected. We are talking about a huge project. If we had the technology this is a project that could boost employment and save our oceans. If we can have drilling platforms why can't we have ocean scrubber platforms? Of course, drilling platforms bring up a product that is in demand. Unless ocean scrubbers supply us with something we need and can sell they will not create any profit. Without profit there is no one to fund such projects in spite of our understanding that our oceans are treasures, and essential to human life on earth.

Related articles

- Ocean life on the brink of mass extinctions: study (empressoftheglobaluniverse.wordpress.com)
- 'Shocking' state of seas threatens mass extinction, say marine experts (redantliberationarmy.wordpress.com)
- You: 'Shocking' state of seas threatens mass extinction, say marine experts (guardian.co.uk)

10.19.11

<u>Oh Snap!</u>

Oh snap!

We will never be able to hold back the rabid plundering of the earth. The hounds have scented money, huge money, and no discussion of H_2O is going to stop them.

According to the tiny bit of the Wall Street Journal I can read without a subscription " [t]he technique of cracking open shale rocks to release oil has spurred hundreds of billions of dollars worth of deals." "Halliburton Co. reported a record $6.5 billion in third quarter revenue."

At a time when we need both money and domestic oil and gas we will be unable to convince anyone that we need to protect our water. Meanwhile in Pennsylvania there is still a town that can light its well water on fire; that has to have fresh water delivered to tanks daily.

I live next to one of the most polluted lakes in America, although it is much cleaner now than even a decade ago. It was contaminated by chemicals used by local industries and then stored near the lake where they leached into the water. I do not trust companies that promise they can do unsafe things safely. They mean well but their records are terrible. Water will get polluted by the chemicals they have to pump into the ground. How far-reaching the effects will be no one can say? There is also a concern that we are totally draining the earth of nonrenewable resources when we should really be slowing down use and stretching out the resources we have.

I picture a giant straw sticking out of our little planet (actually many giant straws) sucking out oil and gas that is trapped between layers of brittle shale rock. This is not the image of a plentiful resource. This is the image of getting every last drop until we are left with an empty planet.

Related articles

- The Truth About Fracking (preview) (scientificamerican.com)
- Shale Boom Reshapes U.S. Energy Industry (online.wsj.com)
- Halliburton CEO: "Phenomenal" shale opportunities outside North America (gcaptain.com)
- Safety First, Fracking Second (scientificamerican.com)

11.7.11

Oil and Water Don't Mix

Will we chose oil or will we chose water? You know what they say, "oil and water don't mix". This is true in science and it is true in the oil business. Wherever there is oil, pristine water resources are placed in a situation of potential jeopardy. This is one key to the opposition of some groups to the Keystone Pipeline Project. The opposition is strong enough that protestors formed a line around the White House and joined hands. I don't know if there were enough people to make it around the entire White House but there were enough to attract Obama's attention. The pipeline decision has been delegated to the State Department, but that could change if Obama responds to the demands of the demonstrators.

NPR says, "In 2010, TransCanada completed a major pipeline – the Keystone – which runs from Alberta to Illinois. The company is now planning a second line, called the Keystone XL, that would run from Alberta to Nebraska with an extension from Oklahoma to the refineries on the Gulf Coast. The map I have included shows both the existing pipeline and the new pipeline which is the center of the current controversy.

Apparently Nebraska stands to feel the greatest environmental impact from the pipeline because the path the pipeline takes has it crossing right through an important source of water. An article that you can find at

http://www.npr.org/2011/11/07/142020522/nebraska-may-play-key-role-in-canada-pipeline-battle.html

says that "Many Nebraskans, including myself [Governor Dave Heineman], support the pipeline, but we are opposed to the route that goes through the Sandhills and over the Ogallala aquifer." "Why would you risk an oil spill or leak over the aquifer when TransCanada already has a pipeline route on the eastern side of Nebraska?"

TransCanada; spokesman, Shawn Howard says opponents are just being unrealistic. "Anybody who looks at this objectively knows that we are decades away from being able to turn off a fossil fuel switch and flip on an alternative energy switch without affecting our quality of life," he says.

While this has a certain ring of truth to it we do perhaps have to be sure to keep our oil away from our water. Paul Krugman in today's NYT's suggests that one reason we cannot break our dependence on fossil fuels and make important investments in alternative energies is because our lawmakers are heavily invested in fossil fuel industries and owe so much to fossil fuel corporations. Mr. Krugman also suggests that the cost of solar energy is falling rapidly and that this alternative way to generate electricity may soon be widely available, especially if it could ever get the support of people who have a vested interest in keeping this source on the down low.

http://www.nytimes.com/2011/11/07/opinion/krug
man-here-comes-solar-energy.html?hp

In addition there is so much desperation behind the
moves being made to separate us from foreign
sources of oil (which I believe almost all of us
would like to see) the oil industry is not, apparently,
going to be held responsible for cleaning up any
environmental damage they do as a result of
technologies like fracking and piping oil from
Canada's oil sands. This is another aspect of this
controversy that is worth protesting as the
consequences may, and probably will, come back to
bite us all. Everything is interconnected.

So how do we balance our desire to be free of
foreign oil, our voracious appetites for cheap
power, our lawmakers' voracious appetites for
profit, and the absolute need to protect our
environment? Let's see if they back off from taking
the pipeline across the Nebraska aquifer and keep
that leg of the pipeline in the eastern half of
Nebraska.

Let's see if "our" government heeds our concerns about hydrofracking. The track record is not good. I expect that, in spite of our objections, TransCanada, will do exactly what they want to do with the blessings of our federal government. We will have to keep trying to be heard and trying to conserve energy.

1.12.12

<u>Dependence on Foreign Oil</u>

I think we all agree that our dependence on foreign oil makes us vulnerable to economic and political manipulation by oil rich countries, most specifically Middle Eastern countries who are experiencing lots of turmoil or who are openly hostile to the United States. Iran's recent threats to close the Strait of Hormuz are a case in point. We also can see that, because of globalization, there is more competition for the oil that is available for sale, which in turn, increases our vulnerability.

There are solutions being offered which are both appealing and unappealing. They offer us the possibility of continuing our current consumption levels for a while and they offer less dependence on foreign oil sources. We can also see the difficulties inherent in almost all of these technologies. Offshore drilling will never be viewed in the same way after the BP oil spill, and nuclear energy, always problematic, has given us new cause for concern after the tsunami in Japan.

The pipeline carrying oil from the oil sands in Canada to the Gulf of Mexico (the Keystone Pipeline) is problematic because of the potential for contaminating the water supply of at least one state (Nebraska) (and don't assume the possibility is so remote that it shouldn't concern us). Hydrofracking uses thousands of gallons of fresh water which is no longer fresh when they are finished with it. Our water resources are more important than our oil resources to our survival on earth. And there will not be any more dinosaurs as far as we know so we will not be encountering a brand new layer of fossil fuels anytime soon. Aren't we planning to leave any resources for future generations? We seem more concerned about not leaving debt for our grandchildren than we do about leaving them with a barren earth.

It has been suggested that by combining all of our energy sources and adding solar, coal, and wind we will be able to meet our needs and become increasingly energy independent. But at the level of individual citizens there has been little impact on our energy sources or our energy consumption levels. If we can afford it we can choose to try solar or wind energy, we can make our homes more energy efficient and buy hybrid or electric cars. However, a larger sector of the population cannot afford to choose any of these energy alternatives. Our power companies are trying to help by offering us energy that comes from a combination of sources, but we need a way to avoid some of the more questionable types of energy acquisition, like oil sands and hydrofracking.

The only suggestions I have to make sound quite frivolous. What if we paid a small tax into an alternative energy fund until there is enough money to buy each family in America some "old school" energy alternatives, like a wood-burning Aga for every kitchen and fireplaces or wood stoves for other rooms in the house? What if we could subsidize a switch to all electric cars? But I guess burning wood is very hard on the environment also and makes for a greater frequency of house fires. Producing electricity requires fuel which leads us back to the same problems we have now. What kind of fuel, how much, and where will it come from? Perhaps this approach would provide enough savings until someone comes up with the next fuel that is as convenient and comfortable as fossil fuels have been.

All right, I know we know we will not be able to give up our fossil fuel addiction right away; life without central heating and cooling is quite uncomfortable. We know we will probably not be able to stop pipelines and fracking, so I guess we will have to just try to get the best possible protections. I just can't help being very nervous about our fresh water supplies. Will the universe save us by giving someone "a eureka moment" to reveal a safe, new alternative energy source? It doesn't look like it will happen right away which leads us back to cobbling together a system that works as well as possible.

4.16.12

<u>Waterworld?</u>

I am relieved to know that I am not the only person worried about water resources on our lovely planet. I was beginning to feel that I was being a weird lefty liberal once again, which is not how I see myself, but, I have been told, is how I may be perceived by others. Writing in the **New York Times** on April 8th, Thomas L. Friedman writes in his op-ed column *The Other Arab Spring* that he sees a tie-in between climate change and political instability like we are experiencing in Northern Africa and the Middle East.

He points out that Yemen is the first country in the world expected to run out of water.

He says, "The Arab awakening was driven not only by political and economic stresses, but, less visibly, by environmental, population and climate stress as well."

He goes on to say, "If climate projections stay on their current path, the drought situation in North Africa and the Middle East is going to get progressively worse…"

"12 of 15 of the world's most water-starved countries according to Nafeez Mosaddeq Ahmed, the executive director of the Institute for Policy, Research and Development in London, writing in the **Beirut Daily Star** in February are (no real surprise) in the Middle East:

Algeria, Libya, Tunisia, Jordan, Qatar, Saudi Arabia, Yemen, Oman, The United Arab Emirates, Kuwait, Bahrain, Israel, Palestine.

He ends with this, "While you may not be interested in climate change, climate change is interested in you.

So here for your enjoyment is a picture of our earth as it is today:

12.10.12

Fracking for Dollars

Should we frack our way to prosperity? We have all these attractive, perfectly sane and nice middle class women with intelligent voices appearing in pastel-colored ads telling us that the gas and oil industry will solve our economic woes. They seem to like to buy ads on news channels hoping, I guess, that this will lend their ads credibility.

But I was watching when the Exxon Valdez spilled its oily cargo in the fairly pristine Alaskan waters. I watched bird after bird, covered with gloppy oil being washed by volunteers and I bet we never saw the many, many critters that died.

I was watching along with the rest of the world when the BP offshore oil well blew up killing workers and filling the Gulf of Mexico, a gulf of beautiful tropical waters and healthy fishing grounds and life-giving wetlands, with a surge of crude oil that went on and on and which they had no idea how to stop. BP oil needed help from almost everyone. We also suspect that that oil spilled in the Gulf of Mexico may still come back to haunt us someday in the future.

We want prosperity, but our water resources are at stake. We can't afford to trust you. We are at least 70% water and we cannot live much longer than three days without water. We cannot afford to frack our way to prosperity until we can truly trust the oil and gas industries.

Humans are flawed and I cannot imagine an endeavor that does not involve accidents that have harmful effects. I think that trust might be having the industry show us a catalogue of accidents that have occurred or could occur and then show us the fixes the industry has developed to deal with each case and tell us what the final impact on people or on the earth, including its water and wildlife, has been or could be.

Even given a process that tries to foresee all possible mistakes and negative outcomes, I don't know if the oil industry will ever be able to regain our trust or if it is even possible to foresee all of the things that could go wrong. Investing in research to find us a new energy source that doesn't rely on fossil fuels might be the best plan of all. Your energy ads are not working, at least not with me. I don't see how sweet spin can trump the images that are stuck in our brains.

12.11.12

<u>The Keystone</u>

I sincerely wish that going forward we didn't need to use fossil fuels ever again but that is not the case. We aren't done with coal, gas, or oil quite yet. It is also essential that we keep trying to set America free from reliance on foreign energy resources, especially oil, because needing to keep other countries happy to insure our oil supply gives them leverage and it gives these same countries potential power to control our response to world events. Also, it is possible that we can use the pipeline as a bargaining chip to help keep the Republicans from insisting on cuts to "entitlements" before they will allow tax rates to go up.

We must accept, however, that we are walking a dangerous line between environmental catastrophes and our energy needs and this dilemma is becoming more and more obvious. Most of us accept that the climate changes we are seeing like the melting ice caps and the rising sea levels and the severe storms can be linked to burning fossil fuels and the levels of CO_2 emissions produced by that chemical process. Using combustion to produce mechanical energy will not work well for us for much longer unless we create domes to live under and move well away from coastlines.

Unfortunately we don't have a great new source of energy waiting in the wings that will provide enough power to meet our energy needs. We have our little collection of problematic alternative energy sources: solar, wind, nuclear, maybe some thermal – each with pluses and minuses. Right now there are more minuses than pluses.

This is why I say we should go ahead with the Keystone Pipeline. I am not really in favor of the pipeline, but I believe they have agreed to change the route so that it doesn't cut across Nebraska's fresh water aquifer. It's practically a done deal and a pipeline is not as bad a risk for our fresh water as drilling offshore or fracturing shale. Sad to say, unless something comes along, we will probably end up doing those also, but let's wait until we're desperate. Let's also keep pushing for the toughest rules we can possibly get to force the energy industry to protect our fresh water (and even our oceans) and to keep CO_2 emissions as low as possible.

1.31.12

<u>How Likely Are Water Wars?</u>

I am not the only person on the planet who worries about water shortages. I have been doing some reading online and there are many reports about the worldwide shortage of fresh water resources. **Scientific American** reports about it and **CNN**, and **Web of Creation** (which may not sound quite as good as the other two sources). Some places never had great reserves of fresh water, places that are obvious like deserts and interior areas of Africa which only have rain for a part of the year and may have droughts that last for several years. We know the American Southwest also is a desert or near desert climate and lacks fresh water resources. If an area is unpopulated the lack of fresh water is not a problem (nature adapts), but as we have spread into areas where fresh water is scarcer, which people have done all over the world, water supplies in these areas become more problematic. Redirecting rivers will no longer do it for us.

The lack of water can mean a lack of food for obvious reason. Crops do not grow without water. When populations try to grow food in low rain or snow environments they must irrigate. To irrigate one takes groundwater and exposes it on top of the earth. It will evaporate and fall again as precipitation, but perhaps not in the same area where it evaporated. It takes 1 ton of water to grow 1 ton of wheat, which makes wheat a water-costly food, says **BBC News**. Many other foods do not require as much water but do not have the appeal of wheat. We may find ourselves having to get used to things like soybeans. Of course meat is also a very water-costly food.

Irrigation, farming, and raising farm animals are also activities that increase the pollution of fresh water. Manufacturing waste pollutes water, or air and therefore water. Retrieving and using fossil fuels also pollutes water in all kinds of ways. Polluted water cannot be used to quench thirst without negative outcomes, including death. Children are especially susceptible to diseases borne in polluted water, especially in poorer countries without water filtration systems and in low-water environments. There are dead water zones even in salt water off many of the coastlines of developed nations worldwide.

Scientists also say that global warming is having an effect on water reserves as snow packs, glaciers and ice caps dwindle in size. The Yellow River in China never used to run dry, then it ran dry for about 15 days a year, and now it is dry for over 200 days a year. It is not the only river that dries up for part of the year when it never did before.

The 10 worst cities in America in terms of available fresh water are not at all surprising. We could almost name them without a list. However, more and more people are moving to these areas. Some populations have grown as much as 20% in the last decade which creates a larger demand for water. They are, as named in an article by **Yahoo Finance**:

1. Los Angeles – Major water supply, Colorado River Basin, Pop. 3,831,868
2. Houston – Major water supply, Jasper Acquifer, 2 Lakes, Pop. 2,257,926
3. Phoenix – Major water supply, Colorado River Basin, Pop. 1,593,659
4. San Antonio – Major water supply – Groundwater – Pop. 1,373,668
5. San Francisco Bay Area – Major water supply, Various, Lake Hetch Hetchy - Pop. Over 1.5m
6. Fort Worth – Major water supply, Multiple – Pop. 727,577
7. Las Vegas – Major water supply, Lake Mead/Colorado River – Pop. 567,000
8. Tucson – Major water supply, Local ground water – Pop. 543,000
9. Atlanta – Major water supply – Lake Lanier, Ga – Pop. 540,922
10. Orlando – Major water supply, Floridan Aquifer – Pop. 235,860

There are spots around the world with water problems similar to these problems of United States cities or some areas with even more pressing needs for fresh water. Will those of us with plentiful supplies of fresh water be expected to share? Will companies privatize our water supplies and sell them to us for big bucks? Will water resources belong to public utilities which give people with plentiful water no choice about sharing water; water might essentially be sold down the grid like electricity. Will these water resources be distributed equally or go to the highest bidder? Will some of us take luxurious showers while others die of thirst?

Oh, we already do this! Will we continue to develop wetlands out of existence although we know how much they contribute to a healthy water cycle? Will we need a Global Water Management Agency? How happy would privileged people be about this? Oh, the protests! Will we learn to control the weather so it will rain where and when we need it to? We can't even desalinate the oceans because we have nowhere to put the brine that is produced as a side product.

How many years of fresh water remain on our beleaguered little planet? What things can we do now to tip the fresh water resources in our favor? Humans are the only species on earth which can manage our water resources. Will we actually do any of those things unless laws are passed to force greater respect for fresh water resources, which for economic reasons, seems unlikely? Perhaps we could all go live in low water areas and leave the great water zones pristine. Then we can all have our water piped in. Humans, for all our capacity to evaluate and recognize problems before they become crises, seem unable to react quickly to take steps to lessen the impact of these problems. It could be a fatal flaw.

10.9.13

Stonewalling While Our Oceans Die

It is getting more and more frustrating to wait for the fruits of winning the 2012 election. The GOP has brought government to a standstill just because they won a majority in the House. They have their obstructionist agenda and they have their "wish" list. Oh boy, they say "we can stop everything until we get a whole array of the items on our "wish" list. In fact, we can get some of them just by stonewalling." These deniers are hurting our nation and they are harming our planet.

Just take the matter of our oceans which, after several huge disastrous leaks and hundreds of smaller leaks, have been inundated with substances which are toxic to life under the seas. Dead oceans will eventually produce a dead planet. All the time these selfish, stubborn people in the GOP are playing mind games trying to get their way, is time that we should be spending in finding ways to save our oceans (besides tackling a few other problems we need to tackle). This may be a far more pressing need than we imagine.

All the time these selfish, stubborn people are obstructing our government we could be building the fleet of space ships we will need to flee our gorgeous planet which we will have to do if we don't figure out how to clean our oceans and keep them clean.

We cannot afford to cling to the status quo, to a global schema which no longer pertains. I hear it as an old blues song: "that old status quo has up and gone away." We are at the end of the fossil fuel era. We are probably at the end of the nuclear option also. We will have to shift to some more primitive, less toxic energy sources (sun, geothermal, wind). By 2050 there will be 9 billion people on this planet living on patches of land between the dead oceans; the very oceans which have always added so much to the glory and splendor of our little planet out at the edge of space, and which have provided so many eons of people with food and other necessary resources. We have always been the little planet that could; now we stand to become the little planet that couldn't.

We need you Southerners and you in the Midwest and wherever you reactionary nuts reside to help save the earth. You need to stop acting stupid and entrenched. Maybe if we give you all federal jobs you will stop trying to bankrupt America and you will get real about the role America needs to play in keeping our Earth healthy. We can't keep fighting these same destructive battles which threaten to undo our nation. We have a lot of problems to solve in the real physical world around us. Get over yourselves!

4.24.14

<u>Don't Bogart That Planet</u>

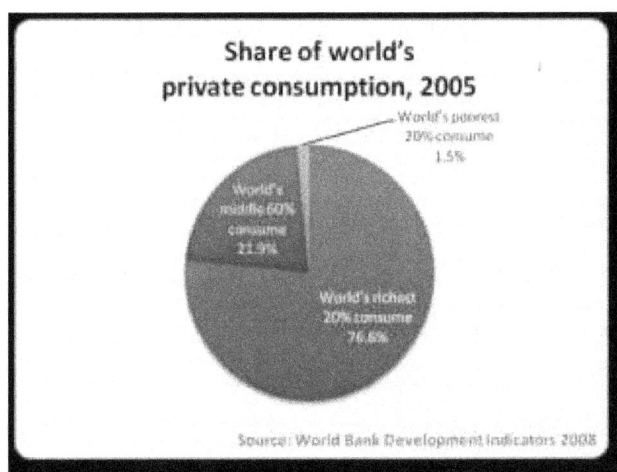

Share of world's private consumption, 2005

World's poorest 20% consume 1.5%

World's middle 60% consume 21.9%

World's richest 20% consume 76.6%

Source: World Bank Development Indicators 2008

I have entertained a faint hope recently that the U.S. has not been quite as greedy about earth's resources as experts say. I was intent on asking that Magic 8 Ball, the internet, if the overpopulation rampant on some continents could be blamed for the stresses we are experiencing on planet Earth (perhaps soon to be planet Reagan).

Alas, the evidence I saw did not support my self-serving hypothesis. Overpopulation is putting stress on the planet, requiring huge quantities of food and water, but the Earth would not be so stressed if developed nations (worst offender, the United States) had not flagrantly plundered and gleefully enjoyed the once plentiful resources of our planet.

I don't think we set out to corner the world's resources, to find and consume them all out of some kind of "mine, all mine" scenario (OK, maybe a few of us did, like the oil companies). We just wanted "the good life" for everyone (in America at least). We were told there were no limits to how high we could climb and we accepted the challenge.

Consumption has been our national pastime in America for decades, and owning nice things, being comfortable, is addictive. But we never said to ourselves, let's use up all the raw materials on this planet. The resources seemed endless and, even though scientists warned us that they weren't, we didn't know how to or had no desire to, stop, so we just denied the evidence. But there are plenty of facts in the form of statistics, out there on that internet which document how we sort of bogarted the planet's resources and that we are still doing it.

In spite of all the deniers, our excesses, however unintended, are starting to have effects on the resources that are still available on this planet to meet the needs of the rapidly growing population (9 billion by 2050). And while I would like to say that the overpopulated nations on our Earth are responsible for the reduction in available resources, they have lived so simply until recently that this is simply not true.

Reducing consumption without reducing use is a costly delusion. If undeveloped countries consumed at the same rate as the US, four complete planets the size of the Earth would be required. People who think that they have a right to such a life are quite mistaken. (This material is taken from the last link listed below.)

Scientists talk about our resources as a two-part concern – the tap and the sink. The tap is what resources are still available, the sink is what the waste of an industrial age is doing to make this an age when the numbers of extinctions of species are growing, and resources are being lost through pollution-by-waste rather than through use.

Fresh water and fertile soils are both being challenged. Phosphates reserves used in fertilizers, necessary to grow the huge amounts of food we will need are dwindling. We haven't yet run out of anything, but it is just a matter of time. How desperate will we get? We are already seeing land grabs and water wars may be closer than we think.

We in America imagine that we will create the world in our image and that everyone on Earth will be brought into the American Dream, rising and rising forever through layers of more and more sophisticated vehicles, homes and appliances. However, it looks like we should really be thinking about simplifying our lives, becoming minimalists, doing that Henry-David-Thoreau-*On-Walden-Pond* thing. I don't believe we have the stomach for this though, and we have Republicans insisting that none of these dire warnings is true and that we can just go on manufacturing, consuming and discarding *ad infinitum*. I wish I believed them, but I just don't and neither do scientists who rely on empirical methodologies as opposed to political ideologies. I don't mean to sound like a socialist, but I am suggesting that Capitalism needs to go on a diet.

So while it may be true that overpopulation will add to our resource problems in the future, the evidence does not support blaming overpopulated nations for the current damages that we are observing to our planet if we are honest with ourselves. We are back to finding a new, really efficient energy source and/or learning how to colonize other planets out in space and neither of those accomplishments can be counted on to manifest in time to avoid the hard choice between some kinds of cut-backs or cultural chaos.

Here are some links to articles I read on the internet:

http://monthlyreview.org/2013/01/01/global-resource-depletion
http://www.worldwatch.org/node/810
http://www.scientificamerican.com/article/american-consumption-habits/
http://public.wsu.edu/~mreed/380American%20Consumption.htm

8.20.14

<u>Sonic Cannons Kill: Save the Oceans!</u>

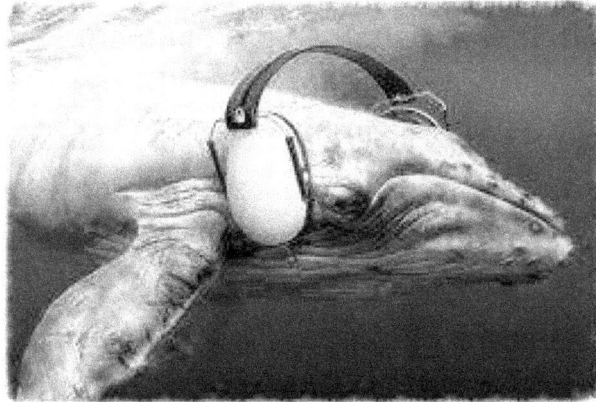

Our oceans are already in a fragile state. Oil spills happen much more often than we know. Tons of waste and garbage pour into our oceans almost constantly. Fertilizers, both organic and inorganic ones which are far more toxic, enter the oceans through run-off from farming. There is plastic in the ocean food chain world-wide. Think of all the debris from the tsunamis in Sri Lanka and Fukushima. Nuclear waste flooded into the sea from that tsunami in Fukushima also and who knows how much nuclear waste is given off by a nuclear sub. It is difficult to believe that there is zero waste from nuclear subs because we know the Law of Conservation of Matter and Energy very well. It's from high school physics.

We hear horror stories about susceptible birds that rely on the ocean for sustenance and we hear more horror stories about the fish and the mammals and all the myriads of other species of plants and animals that live in the ocean. Do we even still have a living coral reef anywhere on earth? We should be working hard every minute to find ways to clean up the oceans. If our oceans sicken and die, we die.

So, although I am usually a fan of Obama I am not at all happy to hear that he has authorized scientists to look for oil reserves in the oceans off our east coast. I suppose this sounds rather benign and that proponents might tell us not to get "our knickers in a twist." But the method used to find oil reserves in the ocean involves firing underwater sonic cannons.

We may wonder how scientists can be sure that firing high decibels of sound pulses under water will not cause earthquakes or tsunamis. But apparently the biggest worry is that these "nerve-wracking" sounds kill the denizens of the ocean. Many, many sea mammals, fish, and other sea critters die from the firing of these cannons (there are some links at the end of this post which give actual numbers.) Animals that navigate by sound may be unable to find their way or may lose their hearing altogether. Every single thing we do has repercussions that ripple outward touching on spheres we never anticipated would be affected.

Perhaps we got Europe to agree to help us sanction Russia by promising to find ways to make up the oil and natural gas they would lose. Whatever the reasons, we should not be doing this kind of exploration in any of earth's oceans (which are all connected to each other). We really need to be very cautious and take a long, long view before we plunder our planet in any way, new or old. We have about reached our plunder limit. Any additional plundering is a big mistake – big! Please put away those sonic cannons and don't drill for oil or gas off the America's east coast or any other coast.

Here are some links to published articles on this topic:

http://www.ibtimes.com/obama-paves-way-east-coast-offshore-oil-exploration-controversial-sonic-cannons-1632726
http://hamptonroads.com/2014/07/obama-approves-sonic-cannon-use-east-coast
http://www.tampabay.com/news/environment/feds-allow-sonic-cannons-on-floridas-east-coast-to-search-for-oil/2189029

Here is a site with a petition you can sign:
http://www.thepetitionsite.com/948/382/904/stop-the-us-from-using-sonic-cannons-in-the-atlantic/

12.10.14

<u>Gray Mountain by John Grisham - Book</u>

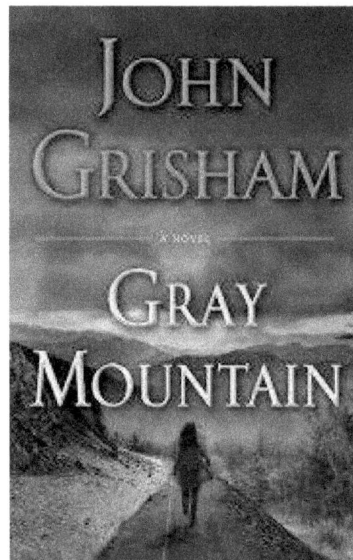

Gray Mountain is not John Grisham's best literary effort but he always delivers the dirt on the "bad actors" amongst us. I watched **The Firm** this weekend (again) and that early book still stands out for both its writing (great suspense) and the theme that has set a fire under John Grisham throughout his career – little people with big moral fiber pitted against power people, money people, corporate people who have lost whatever moral compass they once (or never) consulted.

Gray Mountain pits a few very small characters against a big coal industry that is basically so protected by very powerful people that it can abuse its employees and the people who live in coal country with impunity. These companies get homeowners to sign contracts which allow the coal company to buy their land and they promise huge profits which are seldom delivered. They clear cut mountain tops including valuable hardwoods, strip the mountains layer by layer from the top down until the last vein of coal is gone and then they leave the denuded mountains like sulfurous blemishes on a once beautiful terrain.

They push their leavings into the valleys often killing people who reside there or destroying their homes and they do it without compunction. They can hire the best lawyers so they rarely have to take any financial responsibility for those they harm. They leave a poisonous concoction of coal slurry in dammed-up ponds that seeps slowly into ground water and creates cancer zones and then the coal companies claim that no clear connection can be proven between the cancer and the bad chemicals in the water. This is what John Grisham is great at, exposing greed and its attendant crimes against humanity and the planet.

There is a story line and we do meet some pretty good characters. We meet Samantha, a NYC lawyer, a newbie in a giant law firm who loses her job in the Great Recession due to downsizing. The law firm, in hope of recovery, tells Samantha and the other lawyers if lays off that the firm will pay for their health insurance for one year if they volunteer to work at a nonprofit. Samantha, while not excited about leaving the city she loves, eventually chooses an internship at a Legal Aid office in West Virginia, coal country, the Appalachian Mountains.

Here she meets Matty, her warm, experienced, and highly competent new boss who becomes a friend. She also meets two brothers, one a lawyer, one not, both deeply involved in exposing the sins of the coal companies and for very personal reasons. Samantha doesn't seem like a strong character, but she surprises us as she avoids falling prey to the charms of these two handsome, but not totally ethical brothers. We certainly care about what the coal companies are up to and we end up caring about what Samantha will do also. Whether or not it is great literature, it makes a great read for a winter weekend.

2.1.15

<u>Vote No On Keystone to Save America's Fresh Water</u>

I usually like to write my own articles which are based on a synthesis of what I hear on the news (mostly **MSNBC**), what I read (mostly **The Daily Beast**, **The New York Times**, **Daily Kos**, **New Yorker**), articles I find through Google searches, and my own experiences and insights. But this time I want to talk about the Keystone Pipeline and I don't think anyone cares what I have to say on this subject.

I did start with a premise. My premise is that the very worst problem with the Keystone Pipeline (and there are many problems) is what this pipeline could do to America's freshwater resources in our heartland if the pipeline leaks, which most pipelines do. Have you been following the Yellowstone spills?

The Keystone Pipeline which carries dirty oil from tar sand extractions in Canada, and which offers few benefits to America citizens except for a few permanent jobs and a fair number of temporary jobs, crosses the Ogallala Aquifer in Nebraska. An oil leak that affects this aquifer is a potential disaster when our nation is already challenged to provide enough fresh water to certain regions or the country (California and Texas for example).

So this time I will quote for you from articles that I found in a search of the internet. I always begin with Wikipedia. Although I know it is not the most respected of sources, I look here just to get some quick and dirty information. Then I follow up with more reputable sources, usually a few news items and, hopefully, some more professional articles. Here is what I found in my search, which I will admit was not exhaustive but did turn up some high quality data to back up my premise. The reason I object to the Keystone Pipeline and the reason I cannot stop fighting against it is all about keeping our fresh water supplies safe.

http://en.wikipedia.org/wiki/Ogallala_Aquifer

The **Ogallala Aquifer** is a shallow water table aquifer located beneath the Great Plains in the United States. One of the world's largest aquifers, it underlies an area of approximately 174,000 mi² (450,000 km²) in portions of eight states: (South Dakota, Nebraska, Wyoming, Colorado, Kansas, Oklahoma, New Mexico, and Texas). It was named in 1898 by N.H. Darton from its type locality near the town of Ogallala, Nebraska. The aquifer is part of the **High Plains Aquifer System**, and rests on the Ogallala Formation, which is the principal geologic unit underlying 80% of the High Plains.[1]

About 27 percent of the irrigated land in the United States overlies the aquifer, which yields about 30 percent of the ground water used for irrigation in the United States. Since 1950, agricultural irrigation has reduced the saturated volume of the aquifer by an estimated 9%. Depletion is accelerating, with 2% lost between 2001 and 2009[2][*not in citation given]* alone. Once depleted, the aquifer will take over 6,000 years to replenish naturally through rainfall.[3]

The aquifer system supplies drinking water to 82 percent of the 2.3 million people (1990 census) who live within the boundaries of the High Plains study area.[4]

http://www.washingtonpost.com/blogs/wonkblog/wp/2012/08/10/where-the-worlds-running-out-of-water-in-one-map/

In the United States, aquifers are taking on increasing importance as food production expands and drought becomes a nagging issue. In regions like western Kansas, where farmers don't get enough rain for their crops, they depend on irrigation, using freshwater from the Ogallala Aquifer. That's especially true this year, amid the massive U.S. drought.

About 27 percent of U.S. irrigated farmland depends on the Ogallala aquifer, and it's a key region for livestock, corn, wheat, and soy. But it's slowly getting depleted. In some counties, the water table is dropping by as much as two feet per year. And, as David Biello notes, once the Ogallala gets drained, it would take about 6,000 years to recharge with rainfall.

http://news.nationalgeographic.com/news/2014/08/140819-groundwater-california-drought-aquifers-hidden-crisis/

Aquifers provide us freshwater that makes up for surface water lost from drought-depleted lakes, rivers, and reservoirs. We are drawing down these hidden, mostly nonrenewable groundwater supplies at unsustainable rates in the western United States and in several dry regions globally, threatening our future.

Groundwater comes from aquifers—spongelike gravel and sand-filled underground reservoirs—and we see this water only when it flows from springs and wells. In the United States we rely on this hidden—and shrinking—water supply to meet half our needs, and as drought shrinks surface water in lakes, rivers, and reservoirs, we rely on groundwater from aquifers even more. Some shallow aquifers recharge from surface water, but deeper aquifers contain ancient water locked in the earth by changes in geology thousands or millions of years ago. These aquifers typically cannot recharge, and once this "fossil" water is gone, it is gone forever—potentially changing how and where we can live and grow food, among other things.

Scarce groundwater supplies also are being used for energy. A recent study from CERES, an organization that advocates sustainable business practices, indicated that competition for water by hydraulic fracturing—a water-intensive drilling process for oil and gas known as "fracking"—already occurs in dry regions of the United States. The February report said that more than half of all fracking wells in the U.S. are being drilled in regions experiencing drought, and that more than one-third of the wells are in regions suffering groundwater depletion.

http://www.livescience.com/39625-aquifers.html

Much of the drinking water on which society depends is contained in shallow aquifers. For example, the Ogallala Aquifer — a vast, 174,000 square-mile (450,000 square kilometers) groundwater reservoir — supplies almost one-third of America's agricultural groundwater, and more than 1.8 million people rely on the Ogallala Aquifer for their drinking water.

In addition to groundwater levels, the quality of water in an aquifer can be threatened by saltwater intrusion (a particular problem in coastal areas), biological contaminants such as manure or septic tank discharge, and industrial chemicals such as pesticides or petroleum products. And once an aquifer is contaminated, it's notoriously difficult to remediate.

Of course it is never easy to convince those whose minds are already made up or who are possessed of a collection of opinions that they are unwilling to change. And the legal issue of whether or not a foreign corporation can use *eminent domain* claims to take away the land of American citizens is another area ripe for concern and is something that should make all of us nervous. So, for what it's worth I hope you will read through the articles I found in my search and, of course, if you currently support the Keystone Pipeline, I hope you will change your mind.

5.12.15

<u>Choo Choo Kaboom</u>

The last time the Robber Barons needed to get their products to consumers they built the American rail system. In fact it became a competition as all good capitalist endeavors do. Today's Robber Barons are sending their "oil bomb" trains out on worn out railroad infrastructures and those oil trains are blowing up all over America (and Canada).

Here's something wealthy oil barons could do for all of us and it should be enough in their self-interest that they might actually tackle it. They should voluntarily make safe tracks and safe tanker cars to transport their product. I guess they make so much money that blowing up a town or two is no big deal. I'm guessing that taxpayer money pays for most of the cleanup.

I have seen "oil bomb" trains rolling over my local railroad tracks and this crossing is in a densely populated area. I am sure they are rolling through your community also.

Beneath my text is a link to a YouTube video that documents some of the frightening explosions which have occurred far too frequently across America and will continue to occur because nothing is being done in the short range to change this destructive dynamic.

We are always talking about responsibilities that the Federal government should not take on. Republicans are dying to privatize Social Security. Well if there was ever any expense that should not fall to the government and the American people it is the expenses that result from these "oil bomb" trains and the expenses of remedying the situation.

If we stop allowing "oil bomb" trains to travel across America until sufficient safety features are provided, then I bet those modern-day Robber Barons will update our rail infrastructure right quick.

https://youtu.be/WpXfQMFR_Qs

5.21.15

<u>Sea Sick</u>

We sort of view our oceans, seas, rivers, and lakes as a giant purification system. We can empty any dirty old thing in there and it will come out clean. But we have been fooling ourselves with this science of wishful thinking. When our population was smaller and the waste we emptied into our water was mainly organic, this faux water science sort of worked. But with 7 billion plus people on the planet even organic wastes are taxing our water systems, fresh and salty.

When we were at the peak of the Industrial Age we put heavy toxic chemicals and more complex organics like oil into our water systems routinely. This was done so slyly that we can conclude that businesses were aware of both the true science of their acts and of the moral quicksand they were standing on, but this was the standard procedure in industry at that time and many sins were hidden under our waters.

I'm sure local government leaders and other government people on up the chain were at least peripherally aware of what businesses were doing but industry was so important to an area's economic health that secrets were kept. When our factories left us in The Great Industrial Migration, which began in the 80's and 90's and is still going on, these same governments suddenly found themselves stuck with the toxic waste left behind in local water systems (and in the earth too, in many cases). In some cases local governments were able to hold a business' feet to the fire until they mounted some kind of clean up.

I have seen this whole dynamic play out in my own town which has had the honor of being host to Onondaga Lake, the dirtiest lake in America. Between Allied Chemical and Honeywell dumping heavy metals and other toxins into this small and scenic urban lake and the sewage that overflowed the city water treatment plant whenever it rained, our beloved lake practically glowed in the dark. We could no longer fish or swim in it and if you were boating you shared tales of what would happen to you if you fell in.

There are reasons our planet's surface is three-quarters water. Without a water cycle that repeats predictably life could not exist on our planet. We are 90% water. We only survive a short time without water to drink. Water is life. Water supports marine life which we enjoy eating but which could prove essential to life if we ever do descend into survivalist mode. Primitive man treasured fish because it added variety to their diet and healthy Omega-3's, which they did not have a name for, but which they instinctively knew to be good for them. Even if you don't eat seafood, I'm sure derivatives are in many products you do use.

Now we watch in horror as the petroleum industry expects us to believe that the earth's water system, including the wildlife in it, can absorb oil spill after oil spill and that the negative effects will be purely temporary. But that queasy feeling we get each time a new oil spill sullies another pristine spot is our inner primitive survivalist telling us that we don't believe a single disclaimer from these planet-trashers.

Honeywell may be able to help clean up my community's small lake, but the oil companies cannot clean our oceans. There will be a tipping point eventually when the seas, river, lakes, oceans cannot take the toxic onslaught anymore. I hope we don't wait that long. I applaud efforts to find new energy sources, but we need safeguards for oil drilling and oil delivery that really work and we need them now. (Hint, hint; oil people, stop operating on the cheap.)

Without clean water we will die, rich or poor, we will die.

Here's a few links to lists of oil spills:

http://en.wikipedia.org/wiki/List_of_oil_spills

http://www.itopf.com/knowledge-resources/data-statistics/statistics/

http://www.infoplease.com/ipa/A0001451.html#ixzz3amWW16pK/

Environmental Diary

Part 3

The Heart of the Matter

Conspiracy Theory

Vs.

We Need a Plan

We all, pretty much, can accept that we are living in very partisan times. America seems politically divided almost precisely in half. Republicans are conservative. They have always been partners with business and they believe that their strategies encourage business and job growth. Tax cuts for "job makers" and deregulation are the key economic stimulus approaches they have relied on for decades. They do not worry about protections for workers except in that they think legislating such things is anti-business and therefore hurts the American economy.

This offers insight into why Republicans tend to be climate deniers. It is all about business, all about money; all about keeping America first in both business and power worldwide. They do not care if America has the best benefits or is the most humanitarian nation. These are things a Democrat might value. They feel that economics is the engine of power and therefore all benefits should go to and arise from business. Money is King.

Perhaps Republicans truly do not think the global atmosphere is taking in too much CO_2 and is warming. Perhaps they just don't believe that warming could change climate. It's possible they are all convinced, lockstep style, that human activities have little or no impact on either global warming or climate change, although Americans are rarely monolithic about anything. Of course, they cite scientific studies that back up their position as deniers, but these studies are described as flawed by most climate scientists. They are totally unmoved by evidence that our ice caps are melting right before our eyes. Greenland, actually covered with ice, also seems to be melting.

Democrats, progressives, "greens" have all been talking to Republicans about accepting that environmental concerns need to be a top priority. However, at this particular moment in time, Republicans are defending their own reactionary positions as if their lives and the continuing existence of America both depend on getting their way, absolutely, on all points, no exceptions. This refusal to consider that Democrats have any good ideas, this insistence that Democrats will bankrupt America, and this inability to even consider compromise are actually doing more to undermine America than anything the left has done or wants to do.

Every GOP talking point represents a Republican agenda item and every agenda item is about, at base, the economy – money. Economic health is essential to the survival of a nation, but a society that does not take care of its unfortunate citizens is a heartless and empty culture. A culture that makes sure the rich get richer until the distribution of wealth is extravagantly out of balance will reach a tipping point where there is no more money to move up the chain, or the folks at the bottom must revolt, or both.

We can keep the status quo and head into the future, let the chips fall where they may, and live in the chaos that ensues or we can stop wishing things were like they used to be, hitch up those bootstraps we are so famous for and attempt to face facts so we can hammer out some solutions.

If everyone cannot agree about the math or the science we can at least believe our eyes when they see the photos coming in from our own nation and around the world. Even if extreme weather is not caused by man or by climate change shouldn't we decide what we might want to do if giant storms become the norm for we don't know how long. Will we just take up a collection every time there is a coastal surge, or flood, or wind and water event? As our beaches erode will we keep rebuilding at the water's edge?

 It happens that the nations that belong to the UN held a conference to set an agenda for the 21st century to help us use our money more wisely, to do less damage to the environment (instead of sending it on up to the wealthy who show no inclination to altruism these days.) Republicans react by doing what they do so well right now. They create a conspiracy theory and apparently the masses in America love a good conspiracy theory. In fact, Republicans have not needed many strategies to keep half of America happy. They use lies, propaganda, secret conspiracies, and unrealistic promises. Endless repetition of the aforementioned lies, conspiracies, etc. and quick talking on media seem to do the trick. Half of America falls at their feet.

So the GOP turns Agenda 21 into a UN conspiracy to take over the world – a New World Order (fascist overtones) that gets rid of nations and sidesteps any democratic process by slipping its "sustainability" agenda in through local governments. If your town starts talking about bike paths, community gardens, recapturing rain, consolidating government services, and ending infrastructure support for suburban sprawl areas then Agenda 21 (UN domination) has come to your town. (Are you hearing spooky music?)

But is making our planet into a space that can sustain a population of 9 billion plus humans without using all of earth's resources much too quickly and haphazardly, or running out of fresh water, or burying ourselves in garbage, really a bad thing? Of course, I read my George Orwell, **1984**. I do not relish having "big Brother" manipulate me, not even to do things I might choose to do myself if given the option. All the more reason to be in on designing the future instead of obstructing it.

6.9.11

Too Far Already

*

9.29.11

The Attack on Environmental Protections

*

12.22.11

I Don't Think America is Waiting with Bated Breath

*

3.27.12

The Doomsday Seed Vault – Altruism or Corporate
Plot?

*

5.9.12

Environmental Wars

*

5.10.12

New Data on Global Warming

*

5.17.12

Magic 8 Ball

Asking the Internet

*

5.23.12

UN Agenda 21- A Little Paranoia Break

*

5.31.12

Agenda 21 Again-What's on the Internet?

*

6.12.12

Agenda 21 Revisited

*

6.19.12

Agenda 21-Rio 20 Years Later

*

6.26.12

Rio+20-Quelle Disappointment

*

7.24.12

New Global Warming Math

*

9.27.12

Agenda 21 and Microapartments

*

11.13.12

Cul de Sacs and Agenda 21

*

3.11.13

Bucking the Oil Industry, Part I

*

3.12.13

Bucking the Oil Industry, Part II

*

4.17.13

A Modern Tale- Oligarchy v. Democracy

*

4.27.13

Surprises of Globalization

*

1.9.14

Thank You India

*

1.13.15

Walls-Ideological Walls

*

7.6.15

Middle Ages versus Modern Age?

*

9.29.15

Sustainability

6/9/10

<u>Too Far Already</u>

Government and big business have always been involved in each other's "business". As soon as taxes were collected by government from business these entities became interconnected. After all, governments need money to operate and businesses have more money than most individuals. As soon as regulations began to be written and enforced by the government, businesses had to push back in order to keep as much control as possible. They also wanted to keep as much money as possible. I am not sure we can keep business completely out of government. Politicians are human. Humans like money. Therefore politicians are vulnerable to those who offer them money. We have ethics rules but the temptations are all very great, loopholes are found, ethics rules are broken. So we have individuals dipping into the pot for personal gain, as well as exercising their more altruistic goal of keeping government solvent.

How can we keep big business from exerting undue control over our government and our politicians? The situation is not yet quite as blatant as in **Rollerball**, the movie, in which the corporations are the government. We haven't started turning our trees into fireballs just for the entertainment value.

We don't yet seek ways to turn a sport into a literal "do or die" endgame if a player becomes a heroic symbol of the triumph of the individual over the team/company? Countries do, however, need finances and they need the businesses which generate them. With all the businesses we have lost we can see this loss is making our government less prosperous. Yet we cannot let big business take over our governance regardless of how needy we get. Have things gone so far in America that the corporations are ipso facto in charge of our politicians and therefore our government? Can we afford to make an issue of this right now? Can we afford not to?

Some people believe that big business is already so entwined with government that our government just jumps when big business shouts. Perhaps we are at a crisis point in the relationship between these two entities and perhaps it is good that we are taking a look at the issue now when it might be possible to back up a bit and achieve more balance. We don't want big business to keep pulling up stakes the way they have been. We don't want them to take all their "toys" and go elsewhere, but we don't want to have to let them run amok and take whatever they want.

The environmental impact of big business is probably one of the biggest issues between the

American people and our factories/corporations. If other countries do not put too fine a point on clean air, water, and land, then businesses will leave us because we are making all kinds of environmental rules that are expensive to comply with. Add to this the high wages and corporate taxes and we get the exodus we have been experiencing.

Corporations can be culturally aware and they can police themselves on these and other issues (like product quality and worker salaries and benefits), but it seems that they have not acted on their vulnerability as humans in these areas but have, instead, been concentrating on huge profits. Trying to force big business to consider the planet and their country is like trying to wrestle with an octopus. We can keep trying to appeal to their "better" instincts, but it is unlikely that we will get a positive response when it is so easy to just pull up stakes and move on. It will continue to be a complicated process until the whole world is a level playing field and by then there may be no "fields" left and we many each belong to a "corporate team".

Obviously there are more complicated issues here than just getting rid of "special interests". The skills of some really smart people will be required to guarantee that our needs for a strong economy, a "clean" government, and for the survival of our planet are all being met.

9.29.11

<u>The Attack on Evironmental Protections</u>

Anyone who knows me knows that I am not fond of Republican stances on the issues of the day. I am not even very fond of their belief that the American electorate is so ignorant that basic high school propaganda will bring us all under their sway. In modern-speak they think "[they've] got some moves like Jagger."

So it is no surprise that they are using the economic downturn to turn down the heat on environmental matters. They have placed "clean air, clean water, and clean soil" issues all in their gun sights as mock targets to be shot down. It is quite easy for Conservatives to justify their war on environmental protection because many Republicans say they don't believe in "global warming."

Over and over, they insist that the evidence for global warming has been invented or exaggerated, or any other adjective that will defuse the issue and turn it into a Democratic plot to kill America by making it too difficult to do business in America. This implies that Democrats are so anti-business that they will make up theories; but, I believe Democrats like money, profit and success as well at the next person and would not intentionally sabotage big business.

Republicans tell us *ad nauseum* that our environmental targets and levels have chased business out of America. They assume that if we suspend the rules and let companies "dirty" things up again that these companies will gladly return to do business in America. I think they are wrong. We cannot work for the wages that are paid in Asia and these businesses would have to pick up and move again, which they are unlikely to do.

If there is such a thing as global warming, and I believe there is, even Asia will not be able to operate forever without environmental protections. They may feel that unregulated capitalism is their right because it took so long to come to them, but if they experience an unregulated period equal in length to ours, given their populations and demands, our Earth may not be able to withstand the pollution levels and maintain healthy air, water and soil cycles. Nasty stuff will not stay trapped over Asia. Gases go everywhere, although prevailing winds determine main directions. Seventy-five percent of the Earth is water and water goes everywhere also if you count evaporation-condensation-precipitation cycles. So what happens in Asia won't stay in Asia.

We cannot afford to let Republicans dismantle America's hard-won schedule of environmental protections. I can understand dialing things back a bit while we deal with lost jobs and rising deficits, but we cannot abandon our human position as caretakers of our planet, at least not until we can escape a ruined Earth to live on other planets.

2.22.11

<u>I Don't Think America is Waiting with Bated Breath</u>

Obama's stimulus plan to prod the economy has many aspects. Since we did not get to see if the whole plan would work, the plan was split into its individual components. We didn't get to see if any of those worked either because each separate part of the plan was voted down. Now we have arrived at the last recommendation in the Obama plan, which would allow last year's break on social security taxes to continue for another year and keep extensions to unemployment insurance. It seemed that this part of the plan, on its own, had some chance of being voted into law, until the House linked it with the Keystone Pipeline Project.

Nebraska has a problem with the Keystone Pipeline and, since it puts Nebraska in the position of having two pipelines passing through it, one right over a key water reserve, we are right to support Nebraska on this issue. It should never have been linked to the payroll tax cut or unemployment benefit extensions. The House Republicans knew this was untenable when they did it and to act shocked now is disingenuous. They killed this bill and they did it for political reasons.

I shouldn't really express an opinion on this because I am retired and do not stand to gain any tax relief or unemployment benefits from this plan. Still I can't help having some feelings as I am watching this play out on the news in the week before the holidays. I am not sure the American people want to have this issue as the one that inspires them to say "enough", or that they want to stand and fight on this minor battleground. First of all, it makes us look like charity cases, which we may be right now, but perhaps not so desperate that we have to "take a loan" (so to speak) from Social Security.

This is a very small tax cut, not really a tax cut at all, and some say it may affect Social Security in the future. It is not that working people will not be happy to have a little extra spending money, but we understand that you are lending us our own money to back up consumption and speed up the economy. The extensions of unemployment insurance are harder to give up, but these only help the recently unemployed. So many other groups are not being helped by this. I wish we had gotten grumpier when all the other recommendations in this plan were turned down. We are watching a fake stalemate which is really about the elections and not about helping the American people.

I do hope the Congress passes these two attempts at financial assistance for Americans who need relief. I just hope we remember this represents only one tiny part of the plan to speed up our economy and that the sorry Republicans, who cannot agree to any tax increases on the wealthiest Americans, are not adverse to using this last ditch strategy, which was once part of an organized economic plan, to stage more political theater in Washington. I guess this activity is supposed to help them get elected in 2012 as the watchdogs over American spending. I think we would rather have you keep these pittances and do something more spectacular, risky, and productive that would help us for more than one year and would not put our future at risk.

3.27.12

<u>The Doomsday Seed Vault – Altruism or Corporate Plot?</u>

I started an article about the Doomsday Seed Vault. I thought, how awful, the apocalypse must be near. I thought, how nice, people are looking ahead to after the catastrophic end of the planet earth as we know it. But apparently the seed vault issue does not appear straightforward and altruistic to everyone. I found a fine can of worms and a story that reads like science fiction, but might be true, when the internet introduced me to F. William Engdahl who wrote an article called *Seeds of Destruction, the Hidden Agenda of Genetic Manipulation.*

Remember these words from that Buffalo Springfield song, "For What It's Worth: *Paranoia strikes deep/into your life it will creep/it starts when you're always afraid:*

Reading Mr. Engdahl's article brought these words to mind. I would like to totally discount Engdahl's conspiracy theories about the seed vault because conspiracy theory is what he does and I sort of think of Bill and Melinda Gates, who have helped to fund this project, as philanthropists; but I know how deeply flawed humans are and it is just possible that there is some truth in the points that Engdahl makes. Why is the Bill and Melinda Gates Foundation a partner in this project with The Rockefeller Foundation (which has a Population Control arm) and the US agribusiness giant DuPont, one of the world's largest owners of patented genetically-modified (GMO) plant seeds and related agrichemicals, Syngenta, a Swiss-based GMO company, CGIAR, the global network created by the Rockefeller Foundation to promote its ideal of genetic purity through agricultural change?

Mr. Engdahl sees a number of motives for this particular group of individuals to form a Doomsday Seed Vault in a mountainside on an island off the coast of Norway in the Arctic and none of these motives are the least bit altruistic. They are all designed to take agriculture out of the hands of small farmers and make the world dependent on three or four large agribusiness groups for our food and other plant products like cotton. They do this by making hybrid seeds the only seeds available. There is an element of eugenics in this says Mr. Engdahl, similar to the genetic purity experiments of the Nazis.

Hybrid seeds cannot be collected and replanted in a subsequent season because they are infertile. Thus the farmer must buy new seed from agribusiness providers every season. This eventually forces the small farmer out of business and agribusiness fills the gaps. The small farmers are driven to the cities to be used as cheap labor by manufacturers. (They are all in this together apparently.) Agribusiness currently has its eye on Africa and what Mr. Engdahl calls "the New Green Revolution."

Engdahl also suggests that genetically modified materials are being developed as biological weapons and he suggests that in a recent campaign supposedly designed to prevent tetanus in Mexican women, contraceptives were actually administered along with the vaccines. I wish I could doubt this but America has been guilty of the secret sterilization of at least two populations in our past, and to my once great chagrin, I find that Americans can hardly claim that we have always stuck to the high road. I am not, however, saying that Engdahl claims any connection between the American government and his paranoid or real concerns about the true purpose behind the Doomsday seed vault.

This is certainly better than many science fiction books I have read although I am not prepared to make a judgment about whether or not the facts of the matter prove Mr. Engdahl's theories. After all there is often more than one way to interpret facts. However, I am saying, that knowing what I know, once I have heard that the Doomsday Seed Vault may be another corrupt and greedy plot, I will not be able to totally discount this without compelling evidence to the contrary. A less suspicious part of me still wants to believe these rich folks are just into preserving seeds for future generations. It is also possible that these wealthy people are trying to find a way to feed a ballooning population which will reach 9 billion by 2050.

http://www.bibliotecapleyades.net/ciencia/ciencia_industryweapons31.html

5.9.12

Environmental Wars

Another article sent to me by the Republican in my back yard (TRIMBY) was about global warming. If you watch any of the news channels you probably already realize that many Republicans argue that global warming is not real. They believe that any warming trends we see are just part of the cyclical nature of rising and falling temperatures on earth. It did not help that the study on which Democrats and others based their conclusions about global warming contained made up data.

Republicans argue that Democrats invented global warming to rein in Big Business and, that by calling Big Business to task as the cause of the pollution that in turn caused this "supposed" global warming, they forced businesses out of America and destroyed our hegemony. I cannot imagine why Democrats would want to drive manufacturing out of America. Most Democrats are not socialists, they believe in the economic energy of a capitalist market place. It makes no sense to believe that Democrats would create an imaginary environmental problem, lay it at the feet of Big Business, and deliberately destroy the American economy. This is either deeply paranoid or is an argument being used for political spin by the GOP to tar and feather the Democrats.

As to whether global warming is real, it looks like we do not have enough data to draw a definitive conclusion. There is evidence on each side, none really definitive. The article that the Republican in my back yard sent to me contained a number of graphs that supposedly prove that global warming is not only hogwash, but that the opposite is true, the earth is actually cooling a bit. The problem I had with this article is that the author has a background in statistics which I do not and used quite technical language to make his points many of which rested on whether the data was statistically significant or not. I must admit that my brain turned off a little bit and my eyes started clouding over and I went off into my to-do list for the day. To me it looked as if the temps in the charts actually did trend upward.

On May 3rd, 2012 the article in question appeared in the publication **American Thinker** a conservative online journal. The article "Global Warming Melts Away" by Randall Hoven is a scientific article with many graphs and even a bibliography and is worth checking out.

http://www.americanthinker.com/2012/05/global_warming_melts_away.html#ixzz1u6hJzzF

When the author started talking about linear regression trends for the extent shown that's when my eyes really started to glaze over, but those of you with a good background in statistics can check this out and see if these graphs have merit. I can wait until the evidence is clearer, but I still believe the activities of 7 billion people on our little planet do have some powerful effects on our environment, that these effects are not always positive, and that, in our search for greater comfort, we have not always been careful caretakers of earth.

So TRIMBY and I had an email exchange about these topics of global warming and pollution.

Me: Whether the earth is warming or not, the effects of humans on the earth are huge and not all positive (TRIMBY) (true, so you are admitting that global warming is crap??). (He's a witty guy.) Me: The enormous amount of trash we have stashed all over the planet, the chemicals we have pumped into the air, the soil, and the water are real. He: The earth has much more capability to recover than people understand. That is not to say I (or Republicans, or Conservatives) want to pollute the earth. The reality is, that we do, we will, and there will always be some. Great example is how the ocean "ate" the big spill from the well that exploded in the gulf a few years ago. The spill is gone… Me: The difficulties all this "pollution" presents to the cycles of nature that keep the earth refreshed are real. The exponential growth of the size of human population on the planet is real. He: True in some places, not all: in

Europe, they are having problems with shrinking populations, except in Muslim populations. Me: The eventual exhaustion of the fossil fuels stored under our planet is real He: (true, but it has been greatly under-estimated when it will happen. Further, as it does occur, economics says that we will find alternatives. We are searching for alternatives now and if they are more economical then we will use them. Me: It is not wrong to look for ways we can lessen the negative effects so many people have on the state of our planet. He: Look is fine, force is not. Planning should be more intense rather than less intense. Me: You know I worry that fresh water resources may be our most pressing problem- He: can be in some areas but shortages will raise prices and make people more conscious of the problem Me: and that our reliance on fossil fuels is wreaking havoc on what fresh water we do have (and even on our salt water resources) He: (How so??). Me: I don't understand why people argue that it is harmful to take care of our planet and find ways to keep it healthy. He: It is about balance. You can't have zero pollution, or you can but we would be living in grass huts. It hurts all of us – not just business. You would not be able to afford to eat, sleep, or won a home. It hurts business! Me: That's it? But a polluted earth will hurt business, and it will hurt all living things. We can't listen to corporations when they complain that environmental concerns are hurting their profits (which by the way they are not sharing) He: (they share their profits with the owners of the business, as they should) Me: if ignoring the needs of our planet will eventually kill us. Corporations who are far-sighted would realize that helping to take good care of the earth benefits their bottom line far, far into the future. He: It isn't about being farsighted, the environmental movement has raised awareness of pollution, buy things and really directly influence corporations. This is good. Corporations are about making money – which is good. It they can make money by being environmentally friendly, they will. Me: You and I are both old enough to remember the 60's and early 70"s...the air quality, the quality of the water in the lake, just the general amount of pollution. The environmental movement has been good for everyone. He: Today, corporations know that by being green they will attract more clients, generate more business – make more money...If a corporation is bad, a polluter, then shine the light on it and it will change from public pressure.

Me: As for the fact that our environment is cleaner than it once was, that is because our businesses took themselves off to places where there are fewer laws against pollution and we are now dirtying up other, once pristine locations. He: Again, not really true, there are companies that are owned by China. It is really up to the people there to take control of the country – that is problem with Socialism, Communism, dictatorships...they don't care how the people live. From what I understand they are getting better. Also we are not the world's policemen, we have helped and continue to help this country clean up. If the liberals want to do something about it, publicize it, figure out which companies are the worst and boycott them, put them on the spot…democracy at work. Me: If we don't regulate what businesses can do to the environment they will have no qualms about abusing it for profit.

5.10.12

<u>New Data on Global Warming</u>

Maybe we won't be able to verify global warming by looking at the big picture. Perhaps small local studies by groups who have lived in and observed a specific ecosystem for years will end up being the best evidence for or against global warming. I'm not talking about anecdotal descriptions; I'm talking about data, recorded for decades.

In the **Syracuse Post Standard** on Sunday, May 6, 2012 I found an article with the title *"NY lake's freezing and thawing shows warming trend"* by Mary Esch of The Associated Press which has been picked up by many other newspapers. She spoke with ecologist Colin Beier in Newcomb, NY. He is the lead author of a study that shows that the length of time Wolf Lake in the Adirondack High Peaks region is covered with ice each winter has declined by three weeks since 1975 (so says the label under the picture included with the article). Colin says that "Lake ice doesn't lie. The process of ice formation and lake closure and opening is a straightforward physical process and people have kept records of it for decades."

"The loss of ice cover may change a lake's water chemistry," he says, "and the types of algae and plankton which affect the rest of the food chain."

The article goes on to say that, "Scientists have documented that places like the Adirondacks that are transition zones between temperate hardwood forests and cold-loving spruce-fir forests are especially sensitive to climate change."

"In Vermont's Green Mountains, a 2008 study found the transition zone between maple-beech forest and spruce-fir forest moved 400 feet up the mountain over 43 years, in sync with a 2-degree rise in the area's mean annual temperature."

Small localized studies like these may offer just the data we need to make a clear decision for or against global warming.

http://www.seattlepi.com/news/article/NY-lake-s-freezing-and-thawing-shows-warming-trend-3536827.php

5.17.12

<u>Magic 8 Ball</u>

The internet is a little bit like those Magic 8 Balls we all used to have. Remember them? You asked a question like "Is John in love with me?" or "Will the teacher get the chicken pox?" and then you shook the ball and an answer floated up out of the inky depths. "No way", it might say – or "Probably" might pop up. A search engine works the same way. You type in your question and your chosen search engine wafts your answers through the "net" onto your screen. Just as with the Magic 8 Ball, the way you ask your question is important. Fortunately the internet is much less picky these days. In the old days you often had to resort to an algorithm, which is usually unnecessary in 2012. The answers are often, although not always, better than the ones you got when consulting your Magic 8 Ball.

Anyway I am just talking about Magic 8 Balls to throw TRIMBY (the Republican in my backyard) off the scent because I want to discuss global warming again. (Just kidding.) (TRIMBY is always welcome to speak his mind.) This time the question I asked my search engine was "What percentage of Americans believe global warming is real?" There was actually a range of poll results for this topic depending on when people were polled and how the question was phrased. I also asked the internet to show me whether global warming and climate change mean the same thing, or if there is a difference between the two terms.

Here is the most cogent answer I found in to my second question:

Global Warming is a misnomer...

"The popular term 'global warming' is a misnomer. It implies something uniform, gradual, mainly about temperature, and quite possibly benign. What is happening to global climate is none of these.

It is uneven geographically. Change is rapid compared to ordinary historic rates of climate change, as well as rapid compared to the adjustment times of ecosystems and human society.

It is affecting a wide array of critically important climatic phenomena besides temperature, including precipitation, humidity, soil moisture, atmospheric circulation patterns, storms, snow and ice cover, and ocean currents and upwellings.

And its effects on human well-being are and undoubtedly will remain far more negative than positive.

A more accurate, albeit more cumbersome, label than 'global warming' is 'global climate disruption'.

\\\

Notes: "Global Warming" is commonly used interchangeably with "Climate Change," and other less common synonyms including "global climate disruption," "climate chaos," "global weirding," etc. (ClimateBites offering "climate scrambling")

There is no consensus on the best term. In a famous 1997 memo, Republican messaging consultant Frank Luntz proposed replacing "global warming" with " climate change" because it sounds less alarming, which would make it easier to deflect calls for action. Ironically, climate scientists also favored "climate change," but for a very different reason: the phrase is more accurate because it encompasses all the various predicted changes including regional phenomena, rather than just the average global temperature.

A June 2011 Google search showed "climate change" surpassing "global warming" in search hits by 3-to-2, Holdren's term, "global climate disruption" was far behind.

Bite Source: Presidential Science Adviser John Holdren, quoted in Thomas Friedman, **Hot, Flat and Crowded**, p. 134

http://climatebites.org/climate-communication-metaphors-and-soundbites/terminology/qglobal-warming-is-a-misnomer

Next I asked the question about the percentage of Americans who believe in global warming/climate change and I got a variety of answers:

A **Gallup Poll** summarized in a report on March 11, 2010 (and found on the Gallup Poll website) said that 48% of Americans now believe that the seriousness of global warming is generally exaggerated.

NPR reported on their website on June 21, 2011 that 97% of American scientists feel that climate change is happening. The National Academy of Science (they report) is firmly on board with climate change. Leiserowitz of Yale says (on this same NPR web page), "Most Americans have overwhelming trust in scientists." He continues, "But the public is largely unaware of the consensus (among scientists) because that's not what they're hearing on cable TV or reading in blogs." He goes on, "So far the evidence shows that the more people understand that there is a consensus, the more they tend to believe that climate change is happening, the more they understand that humans are a major contributor, and the more worried they are about it.

In **Rasmussen Reports** for April 9, 2012:

36% say it would be better to invest in fossil fuels than in alternative energy.

The same source asked the following question in a telephone survey with the results below, "Americans recognize more strongly than ever that there is a conflict between economic growth and environmental protection. True or False?"

52 % believe the above statement is true

31% disagree

17% were not sure

Lynn Peeples reporting in the **Huffington Post Green** on February 17, 2012 says:

"If you follow the popular polls, you might think that Americans are growing ever more skeptical about man-made climate change – despite consensus among published climate scientists."

"That's simply not true, Jon Krosnick of Stanford University told an audience of social scientists and cognitive researchers Wednesday in Garrison, NY. He maintained that most Americans do, in fact, believe. He says that, 1) we are not asking the right questions, and 2) legislators are listening to a vocal minority.

He admits that the Gallup and Pew polls have about 50% believing in climate change. He further reports that in his latest poll, where he asked more detailed questions (there is a link to the actual questions in the article) the results suggest that 83% believe.

In addition, he said, not a single state had a majority opinion on the skeptical side.

Perhaps there are more believers than we have been led to "believe". But it seems clear that the jury won't be in on global warming for quite a while. It is a long term phenomenon that we can only verify in hindsight. So I have been schooled and from now on, along with the 97% of scientists in the National

Academy I will use the term climate change, which addresses more short term or local climate changes. None of this is all that helpful in deciding whether DRILL, DRILL, DRILL is an appropriate policy for our government to embrace. Perhaps we can do drill, drill, drill, alternative energy development. Anyway, I am done with this topic for the time being. This represents where I have ended up in my thinking about this topic and I accept that everyone does not feel the same way.

Today in **The Daily Beast** we got this summary of a story from **The Guardian** and we must at least add it to our climate change schema.

8. Australia Hottest in 1,000 Years

A massive new climate study in Australia concluded that the past 60 years have been the hottest in the past thousand for the Australasia region, and that the extreme temperature cannot be explained by natural causes. The study used date from 27 climate indicators, including tree rings, corals, and ice cores, to map the temperature over the past millennium. The climate map was done 3,000 different ways, and concluded with 95 percent accuracy that post-1950 warming was "unprecedented" and exceeded the possibility of an explanation by natural, random variations in climate. (**The Guardian** for May 17, 2012)

5.23.12

<u>UN Agenda 21 – A Little Paranoia Break</u>

The internet is all atwitter – this time about a United Nations document called Agenda 21, and one way or another there may be something to it this time. This 325-page document is a result of a conference in Rio in 1992 by a UN group tasked with coming up with a plan for sustainability. This is a very comprehensive plan for sustainable air, forests, water, trade, and how agriculture, manufacturing, transportation, and all human activity can be included in planned and sustainable communities with planned population growth and health care arrangements. The agenda talks about how to wipe out poverty and equalize incomes. It talks about how to use education to teach sustainability.

Perhaps this plan was intended to be benign, but many are seeing it as anything but. In fact, they see this as a sinister plot to form one global community, regulated from "above", and they aren't referring to a deity. They interpret this as a devious plan to push people off the land and into high density multi-dwelling urban areas of low-cost housing. The land will be returned to a wild state so as to "sustain" those urban areas, the earth, and all the species that remain on the earth. These critics of Agenda 21 interpret the educational goals of the agenda as designed to center the new curricula on only sustainable practices and refer to it as a "dumbing down" process.

This sinister plan is already in place, opponents say, and if you look you will see signs of it operating in your community. It seems that 600 cities in 60 countries have signed onto Agenda 21 and may be devising ways to comply with this agenda that are affecting our lives already. This includes redevelopment of the city core with mixed use buildings, retail on the bottom, apartments above. It includes lots of talk about rail lines and bike lanes. Opponents swear it is the intent of Agenda 21 to kill Democracy, in fact to end all nationalism and leave us with one global system, to end choices, dictate where people will live, what they will do, and how they will transport themselves.

If this scary vision of our future is true, we actually might like to know about it and find ways to prevent such "official" manipulation.

On the other hand, some say that Agenda 21 is a benign plan designed to help us "sustain" our existence on our planet for many generations to come. They say that the United Nations is an amazingly powerless organization. It has no clout. It has little ability to get anything of this scope done at this time. They say that the UN may have written Agenda 21, but no one has looked at it in 20 years and it certainly isn't being systematically implemented.

Even if this isn't a totalitarian plan to move us all over the earth's surface like little pawns until we assume the "proper" configurations deemed appropriate to sustainability, the fact that it exists and sounds so "Orwellian" is enough to give unnecessary ammunition to groups who already feel that environmental concerns were "made up" by liberals to create roadblocks to Capitalism. This will really ramp up the great divide.

Unless it's real, and in that case it just might bring us all together – Foundation vs. Empire.

(Thanks to Trimby for bringing this story to my attention.)

5.31.12

Agenda 21 Again – What's on the Internet?

There is, somewhere on the internet, a video which contains the most comprehensive expressions of the fears some people have expressed about United Nations Agenda 21. It appears to be anything but hysterical. With sober, scholarly people in their suits narrating it seems totally possible that there is a United Nations plan to use sustainability issues for World Domination.

But there are many who feel that this is a fringe "conspiracy theory" and as such is somewhat equivalent to all the hoopla about Area 52. You will have to judge for yourself which side of this debate you will come down on, however, this particular set of beliefs is likely to make environmental issues more contentious than ever, which is difficult to imagine.

Opposite View

Here are some arguments from those who contend that there is no UN conspiracy connected with Agenda 21. This article is from the *Richmond Times-Dispatch* on March 18, 2012 and was written by Rex Springston.

Agenda 21: plot or paranoia?

When the agents of totalitarianism come to crush you, they will do it not with tanks and guns but with electric meters and bike paths.

And your plight, according to their view, will be the work of a United Nations plot for world domination called Agenda 21.

Tea party members and others concerned about Agenda 21 are increasingly popping up at local government meetings to rail against proposals they see as part of the plot.

Among the measures they have tied to Agenda 21, growth plans for Chesterfield and Matthews counties, concerns about rising sea levels along the Middle Peninsula; the Chesapeake Bay cleanup; open-land protections; modern electric meters in homes; and things such as bike paths that are labeled "smart growth" or "sustainable development."

"It is a methodology that has been devised to promote control over resources, the environment and ultimately, people," said Andrew Maggard, a Mathews retiree and avid battler against Agenda 21.

In addition to tea party activists, those opposing Agenda 21 include the John Birch Society, GOP presidential hopeful Newt Gingrich and the Republican National Committee.

Professional planners who have looked into Agenda 21 say the alleged plot is a nonsensical conspiracy theory stemming from long-held fears for the U. N. is bent on ruling the planet under a world government.

"The fact that local governments believe in things like smart growth, livable communities and planning for climate change doesn't mean that local governments are part of a nefarious U. N. plot to take over land-use decisions," said Noah M. Sachs, a University of Richmond law professor and environmental expert.

Agenda 21 – the term means an agenda for the 21ˢᵗ century – is a nonbinding set of U. N. guidelines for protecting the environment, Sachs said. It was ratified in 1992 by more than 170 governments, including the U. S. during the first Bush administration.

"Agenda 21 has been a dead letter for 20 years," Sachs said, "Its recommendations have not been implemented by most governments and the U. S. has largely ignored it.

Those trying to end Agenda 21 – sometimes called "Agenders" – say the federal government pushes the U. N. plan to the local level. Then local officials impose it, often unwittingly, on citizens through measures such as protection of open lands – a move some see as forcing people into dense "human settlement zones" where bikes are preferred over cars.

For more than a decade, few people mentioned Agenda 21. But with the rise of tea parties over the past few years, the issues has arisen with a vengeance.

Donna Holt, director of the Virginia Campaign for Liberty, a tea party group, said she and other Agenders helped defeat Chesterfield's proposed comprehensive growth plan.

"Language in the plan about land and energy conservation, among other things, represented "a blueprint of what I had read coming out of Agenda 21," Holt said.

Another of the Agenders' concerns is "smart meters" in houses – computerized devices that can be read from remote locations. Smart meter, "are by definition, surveillance devices," says a posting on the website Virginians Against UN Agenda 21.

At Dominion Virginia Power, which is just beginning to test smart meters, customers privacy is a "top priority," said spokesman David Botkins, "I never even heard of Agenda 21."

Virginia's Middle Peninsula is a hotbed of Agenda 21 activism, said Lewis L. Lawrence, acting executive director of the region's planning district commission.

Some people see reference's to zoning, comprehensive plans, conservation easements, bike paths, sustainability or smart growth and immediately assert that Agenda 21 is the force behind them, Lawrence said.

"It makes it really hard to have meaningful discussions about what you want to do with your community when 95 percent of the professional language is off-limits because of the supposed nexus to Agenda 21."

Lawrence, whose family goes back to the early 1800's in Gloucester, said he has been accused of being "brainwashed" and "a dupe for the U. N."

Agenders couldn't defeat the Mathews comprehensive plan, but they helped remove what they considered some worrisome references like "sustainable development," said Maggard, the retiree.

Concerns about the environment are affecting property rights and imposing "a draconian form of control over the people," Maggard said, "I cite Agenda 21 as the principal method used to achieve that control."

Much of the Agenders' wrath has been directed toward an Oakland, Calif. Group with the unwieldy name ICLEI-Local Governments for Sustainability. A membership group of local governments, it provides advice on issues such as energy conservation.

Agenders say ICLEI is a conduit through which the U. N. plan moves to local governments. Both Don Knapp, a spokesman for the group, and Holt, the tea party activist, say Agenders were instrumental in getting Albermarle County, James City County and Abingdon to cut their ties to ICLEI.

"It takes very little scrutiny to see that (the U. N. plot) is a complete fiction and paranoia," Knapp said. "It's based on fear. People seem to just keep piling more and more things into this conspiracy theory, and it's absurd."

Some Agenders claim Attorney General Ken Cuccinelli as an ally. Cuccinelli said through a spokesman that he was aware of Agenda 21, adding, "I am concerned about anti-free-market land-use policies that do more harm than good."

Tucker Martin, a spokesman for Gov. Bob McDonnell, said, "Our work is done in coordination with federal and local entities, not the United Nations," McDonnell is committed to protecting property rights, Martin added.

In a January resolution, the Republican National Committee criticized the "destructive and insidious nature" of Agenda 21. And Gingrich says in a You Tube video that the United Nations, through Agenda 21 is "seeking to create an extra-constitutional control over us."

Holt, the activist, predicted more battles over Agenda 21, "I don't see it going away."

http://www2.timesdispatch.com/news/2012/mar/18/tdmain01-agenda-21-plot-or-paranoia-ar-1774490/

6.12.12

<u>Agenda 21 Revisited</u>

People who feel that the UN Agenda 21 (which is supposed to mean agenda for the 21st century) is a plot for global domination and forced environmentalism point to a number of modern trends that they feel prove their point of view. They feel that the foreclosures that have hounded so many Americans and forced them out of their homes were planned by the "sustainability movement" to get people to move away from the suburbs and into high density housing nearer to the cores of our cities. They feel that reverse mortgages are also part of this "sustainability movement" as they will keep the next generation from inheriting properties from their parents.

In addition they cite things like all the recent discussion of bike lanes, green roofs, zero run-off, and other green initiatives as signs that the "sustainability movement" is real, is active and has a deep grassroots agenda.

Then there is the Engdahl analysis of the activities of the Doomsday Seed Bank, in which he says that the seed bank sells farmers hybrid seeds which not capable of reproducing and which therefore must be repurchased every season. Mr. Engdahl believes this is a purposeful procedure designed by the corporate sponsors of the seed bank to drive farmers off their land and into the cities. Is Agenda 21 a plan to manipulate humans according to some grand UN design? Is this not at all the intention of Agenda 21 and simply paranoia on the part of extremist elements in our culture? I'm sure we will continue to hear more about this.

In fact, yesterday, 6/11/2012, I read an article in the **Huffington Post Green** section that says the UN just completed a conference on how to produce a happiness economy to accompany its sustainability agenda. Here's some of what the article had to say:

> *A high-level United Nations meeting on happiness has taken place, marking a significant step towards governments placing wellbeing at the heart of economic progress.*
>
> *...*
>
> *The first of its kind, the meeting took place at UN headquarters in New York on 2 April, 2012, and brought together more than 600 participants from government, academia, business, civil society and spiritual and religious groups.*
>
> *Following the conference, wellbeing is now intended to be at the centre of new sustainable development goals, which are expected to replace the millennium development goals when they expire in 2015.*

"This will add a positive aspiration to improve human wellbeing alongside existing essential goals such as eradication of extreme poverty and universal education," said Mark Williamson, director of Action for Happiness, who attended the meeting.

This is not about being anti-growth," said Williamson, "it's about redefining what we mean by progress. We should be aiming for growth in human happiness. A healthy economy is part of this, but other things are essential too - like vibrant communities and greater equality."

The April meeting was convened by the Himalayan Kingdom of Bhutan, which in the 1970s introduced the concept of gross national happiness (GNH). It began measuring GNH in 2008, looking at factors such as living standards, health, education, culture, good governance, and psychological wellbeing.

In this context, Bhutan describes happiness not as relating to an everyday passing mood, but as "the deep, abiding happiness" that comes from living in harmony with the natural world and with others – that is, from "feeling totally connected with our world."

...

The report, which was published by the Earth Institute and co-edited by leading economist Jeffrey Sachs, states that although the least happy are poorer countries, more important than income are social factors such as supportive relationships, personal freedoms and the absence of corruption.

The report also found that happiness has increased in some countries as living standards have risen, but not in others such as the United States; and mental health is the biggest single factor affecting happiness in any country.

...

Awarded first place in the New Economics Foundation's Happy Planet Index in 2009 and regarded as the 'greenest' country in the world, Costa Rica made primary education free and mandatory in 1870 – before the UK or US – and abolished its army in 1948. In 1970 a network of national parks was set up, protecting nearly 30% of its territory and it now aspires to become one of the first carbon neutral countries.

Will this ameliorate some of the paranoia or make it deeper? Are we talking lobotomy here? Is happiness an attainable goal without legal drugs.

OK, I'm just being ridiculous and cynical, but I can't wait to follow this continuing saga.

http://www.huffingtonpost.com/2012/06/11/united-nations-calls-for-n_1582289.html?ncid=edlinkusaolp00000003

6.19.12

<u>Agenda 21 – Rio 20 Years Later</u>

Agenda 21, which is the subject of much speculation on the internet these days, is recorded in a written document that is 20 years old. The ideas that were agreed to in a conference in 1992 will be the subject of a new conference this week in Rio de Janeiro when over 50,000 people from 193 countries will meet to discuss all aspects of sustainability on Earth once again.

Here is their "mission statement:" "Sustainable development emphasizes a holistic, equitable and far-sighted approach to decision-making at all levels. It emphasizes not just performance but intragenerational and intergenerational equity. It rests on integration and a balanced consideration of social, economic and environmental goals and objectives in both public and private decision-making.

The concept of green economy focuses primarily on the intersection between environment and economy. This recalls the 1992 Rio Conference: the United Nations Conference on Environment and Development."

This conference called optimistically, "The Future We Want" will be held from June 20-22, 2012.

I have written several blogs about the conference of 1992 and there is quite a bit of material about the "sustainability movement" on the internet. A number of people are fearful that we are being nudged into agreeing to plans that sound simple on the surface, but will eventually lead to some overall abdications of our autonomy, our property, and our freedom. These people are being labeled "conspiracy theorists" and perhaps that is what they are.

However, after taking a look at the info graph on this very interesting website I am starting to have a few Twilight Zone moments around this "movement." Maybe we are about to be manipulated to do things that we might agree to do voluntarily if given the total 'vision" of the future that is supposedly being created. Also it looks like societies that are more sophisticated, organized and less energy efficient will be easier targets for these plans than less developed nations, so we're up first:

Here's the link to a video of the info graph:

http://player.vimeo.com/video/43204494"

I will continue to keep an eye on the "sustainability movement," but for now permit me to just say "Yikes!"

6.26.12

<u>Rio +20 – Quelle Disappointment</u>

We have the reports from the Rio+20 "sustainability" conference last week. The conference in Rio in 1992 was extremely productive and resulted in a detailed, organized report called, as we have learned, Agenda 21. Agenda 21 consists of 40 chapters and runs to 325 pages.

It seems as if the economic setbacks of the past 4 years have not created an economy of "happiness" in most of our world, but rather a fearful economy that feels it has to put aside environmental concerns until our economies perk up. So this year's conference produced a document that ran to only 56 pages and mostly referred back to the 1992 document and said "ditto".

Madame Secretary, Hillary Clinton, spoke on the final day of the conference and said, "In the 21st century, the only viable development is sustainable development. The only way to deliver lasting progress for everyone is by preserving our resources and protecting our common environment."

Those who think Agenda 21 is already being implemented at the grassroots level and that this is some kind of grand global plot are unlikely to change their view in spite of the meager outcomes of the Rio+20 Conference. They probably will just say that the grand plan is already in motion and cannot be stopped without vigilant opposition.

Meanwhile our overtaxed environments are not just at the mercy of the 7 billion humans on the planet, but also at the mercy of an economic crisis and a perceived "take-over" that will slowly push earth's populations into configurations not of their own choosing, but which conform to someone's design for what will produce sustainable development on the earth. We don't really like to be pushed around but our tendency to push back may place additional strains on the earth's resources. It would be great if we could all pursue a sustainability agenda together, but highly unlikely.

7.14.12

<u>More New Global Warming Math</u>

Interesting new stuff about global warming this week – **The Daily Beast** calls our attention to a new article published in **Rolling Stone** called *Global Warming's Terrifying New Math*. While it is true that every time we have a warmer-than-usual season people start to get nervous about global warming, it is also true that unusual seasons and more extreme storms do seem to be the new normal. **The Rolling Stone** article opens with some data:

"If the pictures of those towering wildfires in Colorado haven't convined you, or the size of your AC bill this summer, here are some hard numbers about climate change: June broke or tied 3,215 high-temperature records across the United States. That followed the warmest May on record for the Norther Hemisphere – the 327th consecutive month in which the temperature of the entire globe exceeded the 20th-century average, the odds of which occurring by simple chance were 3.7×10^{99}, a number considerably larger than the number of stars in the universe.

Meteorologists reported that this spring was the warmest ever recorded for our nation – in fact it crushed the old record by so much that it represented the "largest temperature departure from average of any season on record." The same week, Saudi authorities reported that it has rained in Mecca despite a temperature of 109 degrees, the hottest downpour in the planet's history."

In the **New York Times**, Monday, July 23, 2012, Paul Krugman wrote *Loading the Climate Dice* in which he gives the following analogy:

How should we think about the relationship between climate change and day-to-day experience? Almost a quarter of a century ago James Hansen, the NASA scientist who did more than anyone to put climate change on the agenda, suggested the analogy of loaded dice. Imagine, he and his associates suggested, representing the probabilities of a hot, average or cold summer by historical standards with a die with two faces painted red, two white and two blue. By the early 21st century, they predicted, it would be as if four of the faces were red, one white, and one blue. Hot summers would become much more frequent, but there would still be cold summers now and then.

And so it has proved. As documented in a new paper by Dr. Hansen and others, cold summers by historical standards will happen, but rarely, while hot summers have in fact become roughly twice as prevalent. And 9 of the 10 hottest years on record have occurred since 2000.

But that's not all: really extreme high temperatures, the kind of thing that used to happen very rarely in the past, have now become fairly common. Think of it as rolling two sixes, which happens less than 3 percent of the time with fair dice but more often when the dice are loaded.

And this rising incidence of extreme events, reflecting the same variability of weather that can obscure the reality of climate change, means that the costs of climate change aren't a distant prospect, decades in the future. On the contrary, they're here, even though so far global temperatures are only about 1 degree Fahrenheit above their historical norms, a small fraction of their eventual rise if we don't act.

Bill McKibben, the author of the article in **Rolling Stone**, tells us three important numbers that we need to keep track of regarding global warming. One is the number 2 degrees Celsius, the second is the number 565 gigatons, and the third is the number 2,795 gigatons.

The significance of the first of these numbers which came out of the Copenhagen Climate Conference in 2009, which produced little else besides this number, was this point contained in the first paragraph of the accord: "it formally recognized "the scientific view that the increase in global temperature should be below two degrees Celsius." It also declared that "we agree that deep cuts in global emissions are required...so as to hold the increase in global temperatures below two degrees Celsius. (We have so far measured an increase of 0.8 degrees Celsius.)

The second number represents this agreement: "Scientists estimate that humans can pour roughly 565 more gigatons of carbon dioxide into the atomosphere by midcentury and still have some reasonable hope of staying below two degrees. (Reasonable, in this case means four chances in five, or somewhat worse odds than playing Russian roulette with a six-shooter.)

Mr. McKibben goes on to say:

"This idea of a global "carbon budget" emerged about a decade ago, as scientists began to calculate how much oil, coal, and gas could still safely be burned. Since we've increased the Earth's temperature by 0.8 degrees so far, we're currently less than halfway to the target. But, in fact, computer models calculate that even if we stopped increasing CO_2 now, the temperature would likely still rise another 0.8 degrees, as previously released carbon continues to overheat the atmosphere This means we're already three-quarters of the way to the two degree target.

As for the third number:

"This number is the scariest of all – one that, for the first time, meshes political and scientific dimensions of our dilemma. It was highlighted last summer by Carbon Tracker Initiative, a team of London financial analysts and environmentalists who published a report in an effort to educate investors about the possible risks that climate change poses to their stock portfolios. The number describes the amount of carbon already contained in proven coal and oil and gas reserves of the fossil-fuel companies, and the countries, (think Venezuela or Kuwait) that act like fossil-fuel companies. In short, it's the fossil fuel we're currently planning to burn. And the key point is that this new number – 2,795 – is higher than 565. Five times higher.

What it Means

It means that this is the moment we must start to switch away from fossil fuels. It means politically and economically it is scary to make this move, but environmentally we must. It means we will become independent of foreign nations that provide us with fossil fuels, but in a totally different way than we are thinking about it right now. It means that unless fossil fuel companies wake up and offer us new ways to make inexpensive energy they will die a horrible death that will be catastrophic for the market place. They must lead the way to non-fossil fuel alternatives or we will have to drive them out of business. They are already in panic mode. Have you counted the number of ads being offered in the media by the fuel industries? We are bombarded constantly by propaganda to make us believe that coal can be clean (will it no longer produce CO_2), that we need to get more fossil fuels from tar sands and fracking the shale laid down by glaciers in prehistory.

It means that, in spite of our economic challenges, we need to find a way to subsidize solar energy to make it much more affordable. We could plaster solar panel all over American housing and let people pay for them with their power bill savings. We need to help everyone switch to hybrid and electric vehicles, again with some kind of subsidies where necessary. We need to put lots of pressure on countries like China and India who are ignoring environmental issues in their understandable goal to raise the standard of living in their respective countries. The planet cannot afford to let them have their fossil fuel moment. These energy companies and energy countries can either continue to be part of the problem which will eventually lead to their demise, or they can be proactive and be part of the solution. The people of the planet beg you to choose the latter approach. Choose it right now, please. Wean us now.

http://www.rollingstone.com/politics/news/global-warmings-terrifying-new-math-20120719

http://www.nytimes.com/2012/07/23/opinion/krugman-loading-the-climate-dice.html?_r=1&hp

9.27.12

<u>Agenda 21 and Micro-apartments</u>

I am thinking that I might have to give Agenda 21 a bit more of my attention. Agenda 21 is the result of a United Nations conference held in 1992. The actual Agenda 21 document is available on the internet as a PDF. It is about 350 pages long but if you read over the 40 chapter headings you will get the drift. I have written about Agenda 21 before, notably during "paranoia week." Conspiracy theorists call Agenda 21 a "plot for global domination and forced environmentalism." The word "sustainability" as in a sustainable life on this planet is the "code" word in Agenda 21 (agenda for the 21st century). Conspiracy theorists are also called agenders. They say that if you start hearing suggestions like roof top urban gardens and bike paths, capturing and reusing rainwater, and idea for improving sustainability (of water, air, soil, trees, etc.) then you are being steer by UN Agenda 21.

Another facet of Agenda 21 is recommendation that government move in the direction of high density housing in urban areas. This week we learned that San Francisco is building micro-apartments which require tenants to live in about 230 square feet.

Looking around the internet it seems that Mayor Bloomberg in NYC asked architects to design apartment in the Kip's Bay area with floor plans between 200 and 300 square feet. These mini apartments are already being built and in fact people are already living in these tiny spaces. After all, there are 7 billion people on the planet. We should be in favor of plans that will sustain the earth's resources.

So when my local community says that they do not intend to extend infrastructure into new areas for developers, when they limit development to infilling areas that already have infrastructure like water, sewers and roads sufficient to handle increased traffic and so on, this is something we should be in favor of. It would seem to be more realistic in terms of our current economic restraints.

However, it becomes unpalatable if it is part of some overall plan to which our local, state, and federal governments have agreed without our knowledge. The idea that someone may be pulling strings from behind the scenes and manipulating us to conform to an agenda that has been kept on the "down low" rubs our American souls the wrong way. We like to have problems explained in clear terms, we like to have input into the solutions that are decided upon, and then we like a role in the implementation of any plans for the future. What bothers us most about this "sustainability movement" is that someone may be moving us around like pawns on a giant chess board. This we would find very unacceptable, if it turns out to be true.

Here's where we point a finger at each of our eyes, adopt a threatening stare, and then point those same two fingers at your eyes. We're watching you!

11.13.12

<u>Cul de sacs and Agenda 21</u>

Quick review - Agenda 21, or the UN Agenda for the 21st century, which was formulated at a conference in Rio de Janeiro in 1992, is all about strategies for sustainability on our planet. There are groups (especially on the Conservative Right) who see this agenda as a sinister move by the UN to take over and create a global government and a global economy subject to the sustainability agenda in Agenda 21. Copies of Agenda 21 are available as a PDF. The document is very detailed and is about 360+ pages long.

Others say that the UN is not powerful enough to impose a world order or any agenda as regards sustainability or any other issue. They say that such green initiatives as roof-top gardens and bike paths are local attempts to cash in on energy savings and grant monies.

However, I keep running into initiatives which seem to be pieces that would perfectly help to complete a puzzle which would look very much like the picture described in Agenda 21. Besides roof gardens, bike paths, etc, the arguments Mr. Engdahl makes about Seed Banks, and especially the Doomsday Seed Bank (see link) which says that farmers are sold hybrid seeds by the seed banks –

hybrid seeds do not generate usable seeds—in each new growing season fresh seeds must be purchased -- and that the point of this is to drive people off the land and into the cities.

We have also discussed "micro-apartments, those new urban domiciles that occupy only 230 – 300 square feet of space (see link). To compare: a 14' x 60' mobile home has about 800 square feet. Now I found a video called "Built to Last" (see embedded link below) advertised on one of my bookmarked pages (a Bing home page I think) which is all about cul-de-sacs and the inefficiency of our current suburban housing model where infrastructure has to be continually extended to serve fewer and fewer houses that are further and further from the city centers, where subdivisions take over farmland and wetlands, and where people burn lots of extra fossil fuels commuting fairly long distances back and forth to work.

The video suggests that we start changing to a more sustainable community model which would involve building small communities that are walkable – communities where schools, grocery stores, etc are within walking distance for everyone living in the community. These "villages" would also snuggle in close to the city center freeing up farmland and shortening commutes.

One difficulty is that this sounds exactly like what we had before suburban sprawl began after WW II. It sounds expensive and wasteful to reverse the trend of the last 60 years, although trends have changed before. It would be more appropriate if we could think of ways to make the homes people already live in more acceptable by changing the infrastructure used to move us and our utilities from place to place. Why doesn't someone work on ways to transport us that don't use fossil fuels – oh wait, we are doing that! Why doesn't someone work on ways to generate energy that doesn't require a complex grid and vulnerable wires – oh wait, we are doing that too! Moving water and sewage are two areas where we have not done a lot of innovation, and most innovation in the other two areas involves technology that many of us cannot afford.

I don't think that we object to trying to sustain the resources available on our little planet in space, except for the people who don't believe that we are plundering and changing our planet.

However, we do love our freedom and we don't like to be manipulated so if it turns out that an Agenda 21 is being imposed on us without our consent that is something we will eventually have to deal with.

Is someone attempting to use propaganda or brainwashing to move us around the chess board, or are all of these approaches just creative problem solving? In the meantime, until we know the answer to that either/or question, some of these individual initiatives, if pursued, may benefit our beautiful Earth and therefore all of Earth's residents (like us). We can deal with the manipulation issues once we have more evidence.

http://youtu.be/VGJt_YXIoJI

3.11.13

Bucking the Oil Industry, Part I

Here's a chapter from Kim Stanley Robinson's book called **2312** which I happened to be reading during this week when I wanted to talk about big places where America is stuck. I was thinking that clean energy was having such a rough time getting the investment and innovation and attention it needs and this is probably because of the oil industry. I realize that this is not too much of a leap, but energy is one of the places where America is stuck. It may also eventually be responsible for the greatest Earth upheavals. Sometimes when you start thinking about a topic the whole rest of your little world starts to chime in with variations on the theme like some great swelling in a symphony or a nexus of culture and ideas and you just find yourself smiling because this kind of mental crescendo occurs so rarely. Of course we are not living in **2312** but Mr. Robinson has a genius for showing us the possible promise and repercussions of our current ideas and behaviors. (From page 303 – 305)

Earth, The Planet of Sadness

When you look at the planet from low orbit, the impact of the Himalayas on Earth's climate seems obvious. It creates the rain shadow to beat all rain shadows, standing athwart the latitude of the trade winds and squeezing all the rain out of them before they head southwest, thus supplying eight of the Earth's mightiest rivers, but also parching everything to the southwest, including Pakistan and Iran, Mesopotamia, Saudi Arabia, even Northern Africa and southern Europe. The dry belt runs more than halfway across the Eurasia-African landmass – a burnt rock landscape, home to the fiery religions that then spread out and torched the rest of the world. Coincidence?

In North Africa the pattern is now disrupted by many big shallow lakes dotting the Sahara and the Sahel. The water has been pumped out of the Mediterranean and deposited in depressions in the desert, often in ancient lake beds. Some of these are as big as the Great Lakes, though much shallower. They're freshwater lakes; the water from the Med has been progressively desalinated on its way inland, and the recovered salts have been bonded with fixatives to make excellent white bricks and roof tiles. White roof tiles covered by translucent photovoltaic film have been used for many older roofs as well; these days when seen from space, cities look like patches of snow.

But clean tech came too late to save Earth from the catastrophes of the early Anthropocene. It was one of the ironies of their time that they could radically change the surfaces of the other planets, but not Earth. The methods they employed in space were almost too crude and violent. Only with the utmost caution could they tinker with anything on Earth, because everything there was so tightly balanced and interwoven. Anything done for good somewhere usually caused ill somewhere else.

This caution about terraforming Earth expressed itself in clots and gouts of sometimes military bickering. Political crosschop led to legal gridlock. Big geoengineering projects were all assumed to contain within them an accident like the Little Ice Age of the 2140's, which was generally said to have caused the death of a billion people. Nothing now could overcome that fear.

Also, for many of Earth's problems, there was simply nothing to be done. The heating and subsequent expansion of the ocean's water – also its acidification – nothing could be done about these. There was no terraforming technique that would help. Some water had been pumped onto the dry basins of North Africa and central Asia, but the capacity was not there to hold very much of the ocean's excess volume. Maintaining the one healthy ice cap remaining to them, high on East Antarctica, was a priority that meant no one was comfortable pumping salt water up there to freeze, as had sometimes been proposed, because if something went wrong and they lost the whole ice cap, it would raise sea level another fifty meters and deal humanity something very like a death blow. So caution was in order, and ultimately it had to be admitted, the new sea level could not be substantially altered. And it was much the same with many of their other problems. The many delicate physical, biological, and legal situations were so tightly knitted together that none of the cosmic engineering they were doing elsewhere in the solar system could be fitted to the needs of the place.

Despite this people tried things. So much more power than ever before was at their command that some felt they could at last begin to overturn Jevons Paradox, which states that the better human technology gets, the more harm we do with it. That painful paradox has never yet failed to manifest itself in human history, but perhaps now was the tipping point – Archimedes' lever brought to bear at last – the moment when they could get something out of their growing powers besides redoubled destruction.

But no one could be sure. They still hung suspended between catastrophe and paradise, spinning bluely in space like some terrible telenovela. Scheherazade was Earth's muse, it seemed; it was just one damn thing after another, always one more cliffhanger, clinging to life and sanity by the skin of one's teeth; and so the spacers kept on coming home, home to home's nightmares, with the Gordian knot tied in their guts.

Good stuff, huh? If we can't even make simple decisions about whether there is such a thing as climate change, about whether or not it is caused by global warming, about whether we should or should not go ahead with the Keystone Pipeline, or about whether hydrofracking will or will not be harmful, how will we ever make the even bigger decisions that probably lie in our real future? Just because we have found rich new sources of fossil fuels, will retrieving these fuels and selling them and using them help our current economy and yet doom Earth to devastating climate changes that will eventually impact on Earth's geography and Earth's inhabitants? We do not know. We are operating in the dark.

Although we have intelligent people providing us with evidence that we still need to find alternatives to fossil fuels, we have others who classify these people as alarmists and who feel that America will lose too much economically and in terms of the balance of power in the world if we do not make use of these valuable resources. They assume that the world's environment is not being changed by man, but is just cycling through climate changes on its own timetable as it always does. And just because these fuel "boosters" happen to own oil companies or invest in oil companies or own businesses that run on oil is no reason to discount them as people who just want to protect their own wealth and power, is it? I think it is clear that the biggest enemy of clean fuel is the fossil fuel industry and that they have no incentive to change the way they operate. In fact, the discovery of new oil and gas resources under America almost dooms any attempt to replace "dirty" fuel with "clean" energy. Our current economy which begs us not to shed more jobs is also a deterrent to changing energy sources at this time. However, CO_2 numbers suggest later may be too late. And there we are. Stuck.

3.12.13

<u>Bucking the Oil Industry, Part II</u>

I had just written myself a note because I had that eureka moment which I talked about in yesterday's post. I suddenly realized that the oil industry has a vested interest in blocking those who want to make clean technology workable. (I know; sometimes I am a little slow.) Anyway, after I wrote that note to myself I went to my mailbox and found the Wall Street Journal for Saturday, March 9th. Tucked into the center was **WSJ. Money**, a special section for Spring, 2013. Inside was an article called *Cleaning Out* with the subtitle: "It wins the hearts of environmentalists, but what does clean tech do for investors? Why some are heading for cover?"

Written by Udayan Gupta this article shows that there has been a decline in investment in clean energy alternatives, and it connects this decline with the new technologies for extracting oil and natural gas which are promising energy independence for America (even perhaps exportable surpluses). Mr. Gupta tells us, "Market analysts at Goldman Sachs, HSBC and UBS are now advising their clients to steer clear of clean tech."

It is fairly obvious that this move is more about profit and less about climate change. We would have to be foolish to ask the oil industry if climate change is real and to expect an objective response. Yet much of the advertising that encourages us to use the newer, cleaner fossil fuels comes from the fossil fuel industry, although they may use that bland-sounding female spokeswoman they like to use to throw us off their scent.

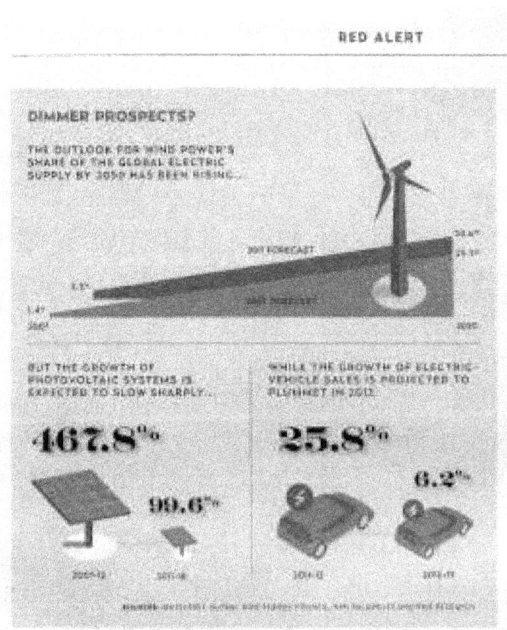

How do you get investors to put money into a technology that is still in the developmental stage when others are raking in the bucks by staying with old energy technology? This would only happen if everyone agreed that continuing to use fossil fuels will continue to release greenhouse gases, which will continue to heat the atmosphere, which will eventually melt the ice caps and bury coastal cities. We don't seem to be having much luck convincing an industry that is making money hand over fist that this scenario is actually in play. Choosing to block a possible negative outcome that may be decades down the road is unlikely when profits are the goal you must shoot for right now.

Mr. Gupta goes on to quote Shad Azimi of Vanterra Capital, "Even worse, the global economic crunch appears to be making it harder for national governments to maintain their funding commitments and tax breaks for clean tech. This development so alarmed climate-change organizations that they put together a group representing $22 trillion in assets and presented a joint letter at a United Nations conference. Their communiqué warned 'further delay in implementing adequately ambitious climate and clean energy policy will increase investment risk for institutional investors and jeopardize the investments and retirement savings of millions of citizens.'"

Abandoning investment in clean fuel, in other words, will not only affect the planet, it will affect citizens who already included such investments in their portfolios. Will we one day wish we never decided to suck the oil out of the depths of our Earth in order to maintain the status quo? Change is difficult and shifting to clean tech is a big change, perhaps too big for an economy invested in fossil fuels and without any certain proof that we must either change or be changed. Still, I am afraid; I am very afraid that what we earthlings really can't afford is to abandon our investments in clean technologies research, development, manufacture and implementation.

Photo credits: Top of page photo is from a Google image search, chart graphic is scanned from the WSJ article mentioned in the text.

4.17.13

AModern Tale – Oligarchy v. Democracy

Let's call this a fairy tale, or a modern tale, and let's imagine a country that stretches from "sea to shining sea". And it is a prosperous country with plenty of jobs that pay pretty well. These jobs are at factories, and at stores that sell the goods from the factories, and warehouses that store the goods from factories, and transportation companies that move goods from the factories.

The people who live in this productive country are doing well. They're not rich but they feel prosperous and they feel confident enough to buy things like vacation homes and second or third cars. They can take trips and support museums and the ballet and symphony orchestras. They are hopeful that the trajectory will move ever upward. Even people who had been left out of prosperity for decades were starting to feel it.

But foul currents are afoot – the factories are making the air unbreatheable and the waters undrinkable. There is smog and pollution and acid rain. All right, these are problems and the people in this country are problem solvers. They decide, we'll pass laws that require factories to create and use procedures that cut down on pollution. But the factory owners aren't happy. It's expensive and cuts into their profits.

The workers need to ask for higher pay to meet rising costs of goods and services. The factory owners get pinched again. A few factories go broke and default on the pensions of the workers. Workers get nervous and ask for higher wages once again. Factories start to leave, wooed by governments with armies of workers who will not need large pay checks or fancy benefit plans. More and more factories leave until the once prosperous country finds that it is no longer an industrial nation. Industry has moved on and left empty buildings everywhere.

It takes a while for things to turn grim but they do start to grim-up a bit. There is no money for beauty; life becomes more practical, even ugly for some. Roads, railroad tracks, runways, trucks, planes, bridges, and all of the country's infrastructure starts to look a little shoddy and neglected. People lose their homes through foreclosure. At first, people think that the factory owners will miss the workers. They were, after all excellent workers. They were their neighbors. So for a while the people wait and find any little thing to do that they can find. But of course the corporations do not move their factories back. And life must move on; a new prosperity must be found.

Obviously this nation is America and while we have been going through all of these changes we have not been getting a lot of help from government. It almost seems that Republicans are working for those same corporations who left us all high and dry. They have been able to have things pretty much their way in spite of the fact that they do not control the Presidency, because they have used their numbers in Congress to block any attempt the elected President has made to stimulate and grow the economy. In fact the word stimulus could hardly be uttered without inspiring lectures and contempt. Conservatives championed a Supreme Court decision called Citizen's United which turns corporations into people, although most of us know that corporations are not people.

Republicans know the American people are not overwhelmingly in favor of what they have touted as "small government" and yet they have insisted that debts and deficits must be cut and have made the government smaller by steering an agenda that has led to cuts and sequesters and more cuts, all of which are working to make the government smaller. They insist that we cut our social insurance programs even though economists tell us that this is not absolutely necessary right now. Once again, as they have their way, government will get smaller.

They have done an excellent job of bullying their agenda into place.

It is interesting to note how much small government would benefit our banks, our financial sector and those corporations who moved out of America. If they can make Americans poor enough, if they can get Americans to let their benefits go, then the ground will be prepared for bringing factories back to America where all those pesky federal, state, and local regulations can go away with all that big government, and unbridled profit will rule the day. Since the labor unions will have also been disemboweled, the cycle of ever-rising wages and benefits will have been broken and workers will have learned their lesson and learned their place and happy oligarchy will ensue.

Do we want our factories back? Perhaps we are done with the Industrial Age. We are already well on our way to finding our way back to a prosperous America that does not have to buy at the company store. Perhaps what we won was our freedom; freedom from big business which likes to use money as muscle to remake America to serve corporate and financial interests. If we don't panic; if we just keep doing what we do, which is to try to live valuable lives that contribute something to our community or to our society or even to the world we live in, then all that creativity and energy will help remake an America that is not just an oligarchy ruled by the wealthy power-brokers. We will become an America that is not a burden on the planet but is, instead, a caretaker of the planet.

We will set ourselves up to be the first country to enter whatever age follows the Industrial Age, perhaps that is the Information Age, or the Robotic Age, or maybe it's the Space Age, or all three. We need to stop these "small government" people from pursuing their reactionary agenda which would take us back where we came from, not to the 1950's everyone supposedly longs for, but to the 1890's which no one except the Republicans even remembers. We need to elect Democrats to Congress in 2014 or we will end up with Grover Norquist's America.

9.27.13

<u>Surprises of Globalization</u>

The admonition of our forefathers that "all men (and women) are created equal" does guide a lot our decisions as Americans and lately seems to keep leading us back to another old adage, that one that says "no good deed goes unpunished". The fact that it seemed wrong to many Americans to enjoy relative prosperity while many others around the world seemed to languish in poverty led to a belief that, although Americans lost all of their jobs, the jobs that were created in places where no boom has gone before (in recent memory) convinced us that this was, in some twisted self-effacing way, a good thing for the whole world in the long run.

Allowing others to make puny wages doing jobs that provided Americans with great incomes could be justified because it would eventually lift up workers around the world, assuage our national guilt, and usher in a future that guaranteed human rights for all. Not that we necessarily had a choice. Globalization happened. Actually, of course, average Americans did not send their jobs to other nations; their jobs were yanked away and bestowed elsewhere. Still it is somewhat comforting to believe that losing our jobs makes us better Americans, adhering to the ideals that formed the basis of our nation and the ideals that people around the world have found admirable and desirable.

I don't think we have been quite as happy with the realities of the road to globalization. It will take many generations, probably, for global economics to raise the standard of living for everyone. In the meantime, Americans are left in a sort of economic backwater, a zone where all but the wealthiest Americans are stuck treading water, and rather brackish water at that.

We don't really want to be in this financial limbo and we may not stay here for long. Hopefully we will find a way up and out, a way back to the prosperity that makes America hum, that calms twitchy Republican plutocrats, and gives us back our optimistic spirit. What we can't know is how long it will take for this to happen, and whether we will be able to pull another rabbit out of our magic hat and find the next thing or things that will take us to a new prosperity. Perhaps on our enforced hiatus from prosperity we will learn to enjoy a bit of languishing, to slow down a bit and embrace a simpler lifestyle that values intangibles like family and friends and leisure and that does not so much rely on collecting more and more stuff, things, objects we never have any time to appreciate.

Must everyone in America have granite countertops and stainless steel appliances? I just saw a photo taken by someone at Reuters and shared on Google+ that shows a Central Asian mother and daughter making cheese. They are squatting in a hut with a straw floor forming perfect mounds of fresh cheese on a wooden board probably getting ready to sell their cheeses at the local market. Obviously the contrast between these two "kitchen" scenarios exposes the distance the world must travel before there is any real economic global equality of opportunity. If we find a way to restore the upward trajectory of our economy the distance among nations will continue to widen or at least maintain its current proportions. However, I don't expect that we will lag behind on purpose waiting for people in other nations to catch up.

In addition, economics is not the only sphere of human activity that has been stirred by globalization. An absolute torrent of hostility has been released, most of it religious in nature between people who adhere to a set of stern religious laws and have practiced this demanding religion since antiquity. So we find ourselves in the midst of a religious firestorm, a maelstrom that was unforeseen by most of us.

If you read science fiction, especially Frank Herbert's **Dune** books, the idea of jihad probably did not come as a total surprise, but still, who knew; not us "ugly" Americans. We did not know that modern communication devices like computers and especially cell phones, and the penchant for tourism that arose with transportation advances and increased prosperity would, just like disturbing a hive of hornets, produce culture shock after culture shock, foment anger and violent reactionary responses that would lead to the threat of terrorism that has arrived on America's (and the rest of the industrialized world's) doorstep and which has become a new fact of life.

Who knew that there are many people who would want to resist globalization, who treasure their traditional lifestyle, their religious isolation and who, once change began to rock their world, awoke to a passion of missionary zeal that Allah requires once the infidel is right in your backyard. Christians ought to understand the often unintentional cruelties of the call to carry a foreign religious mission to "pagans" and "nonbelievers". Many of us did not foresee that what seemed like just simple economic change would resonate through every level of the diverse cultures around the world and make diversity one of the largest issues involved in globalization. Untangling these belief issues and lifestyle issues requires delicacy and time, not strong weapons in the American arsenal. We are spontaneous, well-meaning, earnest, clueless; bulls in the china shop of global human interactions. We are not known for either patience or delicacy.

Now that globalization has begun, it probably can't be stopped unless we go into another "dark" age which seems unlikely. But the globalizations we are experiencing will probably not do away with nations, nor will it probably do away with religions, at least not in any of our lifetimes. Can we wend our ways through the minefields of culture shock and religious intolerance and economic rises and falls to form a more perfect union of the world's nations that could bring to our little planet health and peace? That is the challenge of this particular era of human history.

Will environmental forces trump all of it and drown us in global environmental crisis? We live with that challenge right now. Yikes. I wish I believed that this all arose from our belief that all men are created equal (and perhaps some of it did) but most of this nexus of change arose from greed. Oh well, we are what we are. Surprise! The key words here are delicacy and time.

1.9.14

<u>Thank You India</u>

My mind is wandering to a number of far-flung places and seemingly unrelated thoughts. But, as we decided on Monday, all things are interrelated and even Disney knew this; he said "It's a Small World After All" and he made a whole experience about it and he has made a fortune from it.

So I was remembering that Alanis Morissette song, which I love, "Thank You (India)" because I just heard it over the weekend in a little spiritual film called **The Way** with Martin Sheen in which his son begins a pilgrimage across the Pyrenees to a cathedral in Spain. His son is killed in an unexplained accident so the Dad takes his ashes and walks the pilgrimage. He thinks he does this for his son, and, of course, he does, but he changes his own heart in the process. It is a movie that asks "are you just alive, or are you living."

Then I read TGC Prasad's blog post on wordpress.com in which he expressed his concerns about his country, India. (see link below) India can be quite volatile at times and as India comes to terms with its own income inequality problems (which are even bigger than America's because, although vastly oversimplified, there are so many more people) he worries that chaos will be unavoidable and he hopes this will not be the path India takes to the future (as do we all).

My mind moves on to a story on last night's edition of the national news. Brian Williams is zeroing in on a country the size of Texas in the center of Africa. He is telling me about tribal thuggery in the Central African Republic. Ann Curry talks to an 8 year old young lady whose loses are so great all we can do is cry with her and pray that somehow she makes it through more of this wretched part of her life until she arrives at a life that is not wretched at all. Watching the march of greed, hatred, powerlessness, and some kind of misguided belief that good things will follow from violence gives us all a feeling of the terrible waste of it all and the helpless wish that we had a solution for the Central African Republic, or South Sudan, or any of these storied nations on the continent of Africa.

Which leads me back to America, because I am as self-centered as anyone, and because it is what I know, and leads me to that statement Obama made that caused such rage among the "makers" when he said "you did not make it", and he meant you did not make it all by yourself. And, of course, he was right, in a way. You came up with a business idea. You built it from the bottom up and gradually hired more and more of your fellow Americans. Or perhaps you were wealthy to begin with and your business sprang up as a big enterprise and then just kept getting bigger. You deserve credit for your accomplishment.

But Obama is right also. You couldn't have built your baby, your business without the use of things you don't own like workers and consumers and deliverers, and sellers and warehousers and roads and bridges and railroads and airplanes and power and water. Your factory and your bank balance benefited from a convergence of historical moments which placed you at the very heart of the Industrial Age.

Will you hold on to every last cent even as you watch your once-beloved country subside into the same chaos that Mr. Prasad worries about in his beloved India? We are not usually a volatile nation; it's probably not hot enough here or crowded enough. We will probably just accept our fate as we gradually fall back into the ignorance and hard-scrabble lives of our forebears.

But all I'm saying is that it doesn't have to be that way. You were the builders and you used to share a vision of America with the rest of us "takers", a vision that made America brash and indomitable. Perhaps the vision we have for the future will be different from the one we all had in the Industrial Age, but it can tap back into that creative spirit you used to build the old America and produce a new and even better America, an older and wiser America that will be a proud pioneer in the Global Age.

So thank you India, and Africa and Walt Disney and Alanis Morissette and GE and Carrier Corp.

and Obama and the American workers and on and on and on ad infinitum, which I hope is how long our little blue planet lasts and thrives.

Disclaimer: These lyrics and this you tube video are referenced for your enjoyment. I am not claiming any connection between my ideas and Alanis Morissette. Hearing this song just inspired me and sent my mind on a trip. That is all.

Thank You Lyrics

How 'bout getting off of these antibiotics
How 'bout stopping eating when I'm full up
How 'bout them transparent dangling carrots
How 'bout that ever elusive kudo

Thank you India, thank you terror
Thank you disillusionment
Thank you frailty, thank you consequence
Thank you, thank you, silence

How 'bout me not blaming you for everything
How 'bout me enjoying the moment for once
How 'bout how good it feels to finally forgive you
How 'bout grieving it all one at a time

Thank you India, thank you terror
Thank you disillusionment
Thank you frailty, thank you consequence
Thank you, thank you, silence

The moment I let go of it
Was the moment I got more than I could handle
The moment I jumped off of it
Was the moment I touched down

How 'bout no longer being masochistic
How 'bout remembering your divinity
How 'bout unabashedly bawling your eyes out
How 'bout not equating death with stopping

Thank you India, thank you providence
Thank you disillusionment
Thank you nothingness, thank you clarity
Thank you, thank you, silence

http://www.metrolyrics.com/thank-you-lyrics-alanis-morissette.html

Here's a link to a youtube video:

http://www.youtube.com/watch?v=aSZk4xCTUXA

1.13.15

<u>Walls – Ideological Walls</u>

Totalitarian/Fundamentalist nations have been behind walls for centuries – ostensibly, in each case, to preserve their way of life and their beliefs; but these ideological (as opposed to physical) walls also had the effect of keeping out the "Sodom and Gomorrah" that is, supposedly, the modern world. It may seem that the Iraq and the Afghanistan Wars took the lid off the pot of simmering sectarian hostilities, which many of us were unaware of, thus unleashing both the hate and the hope we are currently pulled and pushed between. I guess we thought that Islam was fairly monolithic, which, obviously, it is not, any more than Christianity is monolithic or Buddhism or any other of the world's religions. When we think about the "hatred" that some Muslim fundamentalists express for Western nations (and/or all non-Muslim people) we tend to blame the current zeal on our wars with Iraq and Afghanistan. But the Muslim world had already decided that Western culture was depraved and was in violation of the Prophet Mohammed's teachings and therefore the Muslim faith.

They had already made this clear, in no uncertain terms, on 9/11. They attacked us before we ever went to war, although I think that lately we tend to confuse the order of events. (I still think going into

Iraq was the wrong choice, except that perhaps someone knew that Saddam Hussein was the weakest link.) Our defense of America after these attacks may have widened and deepened the ranks of those who see Europe and America as enemies, but, on 9/11 when the war began in earnest, factions in Muslim nations were already determined to keep the evil, sinful modern world out, to eventually undo progress that, to fundamentalists, appears antithetical to Muslim beliefs and practices. And these fundamentalists are probably correct in a way. No religion has been unscathed by the cults of the individual and of progress as practiced by modern Western nations.

One would think that jet airplanes or TV would have done the trick of making the ideological walls crumble. However, I think what finally brought down the walls between antiquity and the modern world was wireless technology (which may have become more prevalent with our troops in the area) – internet and cell phones. Visitors arriving on jets could be controlled by authoritarian leaders. TV broadcasts could easily be blocked and rebroadcast locally where content could be controlled by those same totalitarian heads of state. But how do ideological walls stop air, molecules, or electrons? Only North Korea, which carefully controls internet access and access to hardware truly has been able to maintain its walls against this "cellular" assault.

War or no war, our clash of cultures was already an historical inevitability. I always assumed modernity would win, but given our environmental and

climate change issues, the need to decrease our
carbon footprint on the plant may give antiquity an
advantage. Can we keep our human rights
advances if we enter a post-industrial, post – fossil
fuel age? Can Muslims keep the faith if they
modernize? Will those ideological walls disappear
or will the people of Earth turn their bare faces to
the stars from a planet that has retreated wholly
behind those walls to return to more primitive
times?

The only certain things are that change is
excruciating and outcomes are unpredictable but
without change there is no such thing as hope.

7.6.15

<u>Middle Ages versus New Modern Age?</u>

Sunday morning, July 5, 2015, in **The Daily Beast** Joel Kotkin wrote an article entitled "Green Pope Goes Medieval on Planet" in which he projected that if we did what the Pope proposed that we should do for the environment it would move us back into an essentially medieval society.
I am always fearing that if Republicans (and corporations) have their way with America we will become of society of nobles and serfs, except most serfs will not be farmers this time around; they will work in factories. These factories will operate without regulations or rules making the lives of American workers as grim and empty of self-fulfillment as we imagine Chinese jobs to be in factories run by a totalitarian state.

Kotkin objects to the "green movement" plan to shove people into dense urban environments, to do away with suburbs which waste resources and require extended and costly infrastructure builds for water, and sewage. Suburban living involves transportation to and from jobs which are usually not located in suburban neighborhoods. Commuting creates pollution and sprawl is unnecessary and self-indulgent and unsustainable, say those in this green movement (although I don't believe that there is one unified green movement at this time).

Kotkin believes, as most Republicans do, that this is an overreaction to unproven claims about global warming and climate change and even to overpopulation. He objects to this as cattle-prodding us towards losing our personal space in close-packed urban centers, although this move would return wide swaths of suburban land back into farmland or natural habitats to help us feed the coming multitudes. Interestingly enough the Republican plan to turn most of us into cheap labor "fodder" in order to keep the world supplied with goods (although who will be able to buy them) and to keep that conveyor belt carrying money to the corporate elite will also function very well to answer the needs of the "green movement" at least in the sense that most of us will have a much smaller "carbon footprint". The green movement will not be happy, however, with the extravagant burning of fossil fuels. Will so many people being forced to live simple, feudal, therefore medieval lives balance out the fossil fuel assault on the earth and therefore slow down global warming?

Well I hate to see us go backwards at all. The Middle Ages, after all, were also known as the Dark Ages; years with very little learning, art, music, and philosophical thought illuminating the lives of most people. They were also cold in winter and hot in summer and people were often poorly clothed and poorly fed. Mercifully their lives were often short.

Do we have to go back there? Not if we have the foresight to plan and the fortitude to hold out and fight for the best deal that "commoners" can negotiate. If makes so much sense to take care of the earth that we should not have to argue about whether the damage has affected the planet or not. It is clear that we have not been salting away earthy gifts because future generations might need these gifts also. It make sense to be smart and not use and abuse the earth even if you choose not to listen to what science has to say.

We should be able to keep our books and remain a well-educated public, to keep our creative faculties intact, to maintain a cultural mélange and keep an enlightened respect for human rights and still find ways to be more minimalist; to live lightly on the earth. If the choices are to either husband the earth's resources or to go out in a somewhat smoggy blaze of glory (for the wealthy) and grim labor for the rest of us, then I would rather try the plan that the "green movement" has hatched for sustainability.

Republicans believe that we do not need to follow a plan to live a sustainable life on earth. They believe that no matter how we plunder the earth, it will be fine because God knows what he's doing. If it seems to some of us as if God sort of lets affairs on earth muddle along without divine interference then we had better pick the greens over the reds (Republicans) and allow ourselves to be voluntarily herded into whatever configuration will help us live on earth for the foreseeable future as long as they

only try to move our bodies, but leave our minds
free.

http://www.thedailybeast.com/articles/2015/07/05/g
reen-pope-goes-medieval-on-planet.html

Environmental Diary

Part 4

Attempt to Solve

Environmental Problems

And

Talking About

Saving Nature

People have been thinking about, creating products for, and acting on their own concerns about the environment. Perhaps this process has not been at all organized, has been sort of herky-jerky, but stuff is happening. This diary does not give much coverage of what is happening with solar and wind technologies but these have become viable, mainstream industries even though Republicans still love to indict Obama by shouting out "Solyndra" anytime alternative (non-fossil fuel) energies are mentioned. Solyndra may have been a big fail, but there have been many successes also. I now get my electrical energy from a solar/wind company. When I go to visit friends in northern New York windmills keep company with the corn in the fields.

Since nuclear waste has such a long half-life and must be stored securely for centuries. We have discussed catapulting nuclear waste into space but we don't have the technology to do this and if we one day can head out into space we don't want to have to dodge our own nuclear garbage.

In Europe they are building something called tokamaks to provide clean energy. There are no tokamaks on line yet so we will have to wait to see what happens, but there is a diary entry about tokamaks. A couple in America has devised smart (strong) solar panel system for our roads with all the infrastructure hidden neatly beneath but easily accessible. They are already manufacturing these panels and some people are buying. This will give us some data about how these roads perform. Will they work in snowy areas like mine? That point was not addressed on the site I visited (link provided in diary). Aren't they prohibitively expensive right now? Still, ingenious, and a sign that Americans are putting their creative minds to work on creating new and useful environmental products.

What do asteroids have to do with alternative energy? Higgs Boson? Trees? Canada Geese? Desert Dwellers? Globalization? The connections are more tenuous, but we know that everything is interconnected. Asteroids can be mined, although not for fossil fuels I'm thinking. I can't picture a dinosaur riding an asteroid through space. The Higgs Boson might be as important at some point as Einstein's $E=mc^2$ proved to be. Trees – no brainer- trees use the CO_2 that we can't use and turn it back into O^2 which we can use. We just can't have too many trees. And Canada Geese are telling us that we are perhaps covering over our wet lands excessively and destroying habitats that are important in the grand scheme of things.

Desert dwellers are being affected more than you would think by climate change. Here is the link to a stunning article from the New York Times about people in the desert areas of NW China. The next day's times continues this story.

http://www.nytimes.com/interactive/2016/10/25/world/asia/china-climate-change-resettlement.html?hp&action=click&pgtype=Home

Desert dwellers in the Middle East are also experiencing political upheaval which is affecting the entire planet. While ISIS and other terrorist groups consume our concern and our psychic energies, we are unable to concentrate on environmental issues that could well make our lives far more miserable than ISIS ever could.

We should be forming think tanks, offering financial incentives and prizes, sharing ideas, and even designing our own overall sustainability plans if we don't like the one the UN nations came up with. But with a nation half full of deniers and those deniers perhaps representing the wealthiest half, we cannot move forward and that other half of our nation wishes to insure that we don't move backward, slurping up the entire fuel reserves of our planet with no thought to the future. You would think Republicans, usually so careful in economics, would be equally parsimonious with our natural resources.

3.26.11

The First Law of Thermodynamics

*

3.8.12

Attack from Outer Space

*

10.27.12

Protecting the Grid

*

11.14.12

Off the Energy Grid

*

8.13.13

This is the Droid You're Looking For

*

9.5.13

Going Bigger, Much Bigger

*

3.10.14

Saving Earth – Tokamaks

*

6.11.14

Great Minds United Might Save All

*

6.17.14

Appealing to People in Ancient Lands

*

2.16.15

Here Comes the Future of Infrastructure

*

Saving Nature

*

4.13.11

Some Green Please

*

7.23.14

How Canada Geese Became Pests

*

9.25.14

Apparently Greed Makes You Stupid

*

4.12.16

Global Concerns

Also on You Tube at:

https://www.youtube.com/watch?v=agsRiiBsFJc&t=3s

*

5.1.16

Desert Dwellers and Polar Ice Caps

3.26.11

<u>The First Law of Thermodynamics</u>

I experienced a deep personal satisfaction when I finally understood the law of conservation of matter and energy (also called the 1st Law of Thermodynamics). I learned this law in a rote fashion in high school physics but I did not really understand it until I was an adult. It is sort of an elegant little law which apparently engenders no end of complicated experimentation and theorizing among physicists and chemists, and others.

The law simply states that matter and energy are neither created nor destroyed but are changed in state (e.g., solid, liquid, gas). It finally occurred to me that what this law says is that there is a finite amount of energy in the universe and that new energy is not being created somewhere outside the system and added into it. It also implies that there is a constant state of flux in the universe in which matter may change its state, or matter can become energy and energy can become matter. This helped me understand algebraic equations (why what you have on one side of the equal sign must equal what you have on the other side of the equal sign), chemical equations (its why they are balanced), the water cycle, photosynthesis, heating and cooling, atomic energy, $E=mc2$, ashes to ashes, dust to dust (with room for argument), and that very male phrase which keeps cropping up lately "zero-sum".

Eureka! Matter does not disappear; it just changes form. It suggests that there is order in the universe, a sort of cause/effect, although often on a scale so huge or so tiny that it is sometimes impossible to predict. We even have chaos theory which says that even in chaos there is order. And string theory which looks for similarities in the order of things (I think).

Even though we understand how this all works, we cannot always control it. $E=mc2$ says something with a tiny mass can exert an amount of energy multiplied by the speed of light squared, which is a huge number. This is the basis for nuclear energy. When we can control the reaction and turn it into electricity, nuclear energy is very useful, but we know we are putting a human leash on potentially huge forces and we know the rules for keeping these huge forces under control and we know the dangers if the rules are circumvented or cannot be followed.

If it were just a matter of creating energy that would be fine, but the only substances we can use to create this energy at the present time are substances that are unstable, radioactive, and injurious to humans above certain levels. The radioactivity of these substances is also long-lasting, it can be diluted, but it doesn't subside for a long, long time. This is sort of equivalent to the Hindenburg disaster when they used hydrogen to make a dirigible lighter than air.

Hydrogen is a very active element. It forms bonds easily with other substances and differing amounts of energy can be released. In this case a spark caused a bond that produced an explosion and a spectacular and horrifying accident. Using helium solves the problem, because helium is inert and usually nonreactive. That's what we need, a way to produce nuclear energy that does not have the side effect of producing dangerous levels of radioactive material and tons of long-lasting radioactive waste. I'm obviously no scientist. Maybe this cannot be done. Maybe it can be done and it will be the answer to the future. $E=mc^2$ – we know there is something there.

3.8.12

<u>Attack From Outer Space</u>

I guess since we can't go to space right now, space has decided to come to us. We are in a very active solar storm period. For the last ten years things have been fairly quiet on the sun, but lately the sun has grown stormy. Today a really giant solar flare resulting from an intense solar storm will zip across space to hit us here on earth. Fortunately solar flares have been hitting earth for centuries and the effects are not too terrible. We might see, experts tell us, some disruption of electrical grids and navigational equipment. Others will observe the effects of these solar flares for us. It will be interesting or perhaps anticlimactic to hear what they report. The Northern Lights generated by solar flare activity have been spectacular in some places. I have always wanted to see the Northern Lights but since I live in a city with lots of cloud cover I have never had the pleasure.

We are also being warned as recently as yesterday that early next year (no real date yet) an asteroid will pass so close to earth that it will pass between our satellites and the earth. We are also being assured that this asteroid should not damage the earth in any way. If you are superstitious however, the date given for our close encounter with an asteroid and the date given in the Mayan calendar for the end of our human tenure on this planet are quite close. I believe the Mayan calendar says that on December 21st, 2012 the earth's poles will shift and we will be ejected into space. If so we may not have to be concerned about the approaching asteroid at all; we won't be here. If not, yikes!

We are also informed that scientists may have located the elusive "God Particle" or Higgs Boson. The term "The God particle" was coined by physicist Leon Lederman in 1993. In his book by that name, he said, "If the Universe is the answer, what is the question?" (That's how physicists talk.)

Scientists in Chicago and Geneva are competing to be the first to find evidence that the particle actually exists. In Geneva, and previously in Chicago, "scientists have collided protons with antiprotons in an experiment to discover how matter gathered mass after the Big Bang, 13 million years ago." "Higgs boson is said to have made stars, planets and life possible by giving mass to most elementary particles." Scientists announced this week that they have glimpsed it and will know the answer by the end of 2012. I don't want to wish the year away, but how will we be able to stand the suspense of waiting for all these end of 2012 events. Bazinga!

http://www.foxnews.com/scitech/2012/03/06/asteroid-to-buzz-by-earth-next-year-time-to-call-insurance-agents/

http://www.mirror.co.uk/news/technology-science/science/scentists-close-in-on-finding-the-god-particle-755034

10.27.12

<u>Protecting the Grid</u>

Recently we were warned once again about possible terrorist attacks on the grid and the internet. Shutting off our electricity in almost any season would be uncomfortable and difficult but with winter coming it could be deadly. Given that our grid has become increasingly centralized in interdependent districts which, if disabled would affect huge areas of the United States, the possibility of outages caused by hacking or similar tactics is intimidating and could have very serious outcomes. Similarly disruptions to the internet, on which we have grown increasingly dependent, would be equally daunting.

The dangers are spelled out in an article entitled *DOD official: Vulnerability of U. S. electrical grid is a dire concern*, by Dan Merica.

> *Speaking candidly at the Aspen Security Forum, one defense department official expressed great concern about the possibility of a terrorist attack on the U.S. electric grid that would cause a "long term, large scale outage."*

> *Paul Stockton, assistant secretary for Homeland Defense and Americas' Security Affairs at the Department of Defense, said such an attack would affect critical defense infrastructure at home and abroad – a thought that Stockton said was keeping him up at night.*

What are our options? How do we defend ourselves against cyberattacks that would be designed to disrupt the American economy and could possibly be quite effective in accomplishing such a goal? Decentralization might be one obvious choice. Divide the large grid districts into the smaller districts we used to have. This would make it difficult, however, to direct power to where it was most needed. So if giving up huge grid control centers is not desirable then we need some other solution. So we need an interconnected grid that operates like a parallel circuit rather than a series circuit, in other words, it is connected but also separate. And of course, we might be able to section off parts of the internet, but this also seems counterproductive to me, unless there is a way to minimize the separations for legitimate users.

Discussions for making a smart grid are not designed to protect the grid from terrorist attacks but they are designed to make the grid easier to maintain. Using smart grid methods is problematic because there are privacy issues. If there is a meter that can communicate to the grid from inside our homes without being read by a physical person, people worry that other info could also be gathered in this manner without our knowledge. The following explanation is from the energy.gov web site:

> Smart grid" generally refers to a class of technology people are using to bring utility electricity delivery systems into the 21st century, using computer-based remote control and automation. These systems are made possible by two-way communication technology and computer processing that has been used for decades in other industries. They are beginning to be used

on electricity networks, from the power plants and wind farms all the way to the consumers of electricity in homes and businesses. They offer many benefits to utilities and consumers -- mostly seen in big improvements in energy efficiency on the electricity grid and in the energy users' homes and offices.

For a century, utility companies have had to send workers out to gather much of the data needed to provide electricity. The workers read meters, look for broken equipment and measure voltage, for example. Most of the devices utilities use to deliver electricity have yet to be automated and computerized. Now, many options and products are being made available to the electricity industry to modernize it.

The "grid" amounts to the networks that carry electricity from the plants where it is generated to consumers. The grid includes wires, substations, transformers, switches and much more.

Much in the way that a "smart" phone these days means a phone with a computer in it, smart grid means "computerizing" the electric utility grid. It includes adding two-way digital communication technology to devices associated with the grid. Each device on the network can be given sensors to gather data (power meters, voltage sensors, fault detectors, etc.), plus two-way digital communication between the device in the field and the utility's network operations center. A key feature of the smart grid is automation technology that lets the utility adjust and

control each individual device or millions
of devices from a central location.

Wikipedia summarizes the arguments for decentralizing the grid into Micro grids:

> *Decentralization of the power transmission distribution system is vital to the success and reliability of this system. Currently the system is reliant upon relatively few generation stations. This makes current systems susceptible to impact from failures not within said area. <u>Micro grids</u> would have local power generation, and allow smaller grid areas to be separated from the rest of the grid if a failure were to occur. Furthermore, micro grid systems could help power each other if needed. Generation within a micro grid could be a downsized industrial generator or several smaller systems such as photo-voltaic systems, or wind generation. When combined with Smart Grid technology, electricity could be better controlled and distributed, and more efficient.*

These would be great areas on which enterprising young internet wizards could focus their attentions. Designing useful national security approaches for both our electrical grid and for any computer sites that need to be absolutely "unhackable" would be worth the offer of a government prize or prizes to the winning creator or team of creators. However, the real problem, besides a need for someone to create the technology, would be finding the money to implement the plan once it is designed. Money, in fact, will be a consistent litany keeping us from the future we need to build and the availability of funds will rely on rebuilding our economy.

http://security.blogs.cnn.com/2012/07/27/dod-official-vulnerability-of-u-s-electrical-grid-is-a-dire-concern/

http://energy.gov/oe/technology-development/smart-grid

http://en.wikipedia.org/wiki/Electrical_grid

11.14.12

<u>Off the Energy Grid</u>

When I went to the Road Runner home page there was an ad on my page that accused President Obama of supporting policies that are causing our utility bills to rise. It was one of those mean-spirited conservative tirades that could not just state that heat and electric bills are rising, but which had to place the blame for this on some kind of government conspiracy. The video suggests that the rise in costs for energy has something to do with environmental protection but he never mentions climate change and the need to control CO_2 emissions. He calls himself a patriot and he lets us know in no uncertain terms that he feels that our government, especially our government as led by Democrats and Obama, is deliberately making not only mainstream power sources more and more expensive but they are making sure we don't have access to solar power and wind power by making the costs of buying this technology prohibitive.

Now, if you put aside all the hate rhetoric and conspiracy theory that, for me, clouds the message that Frank Bates has to deliver and you just listen to the practical information he has to give us, when he finally gets around to it, then the whole video may have some value. He says he will tell you how to find the components for and how to build inexpensive solar panels and wind power equipment and to eventually get completely "off the grid". He says his video has been banned by Google, but I doubt that very much. If it was it was because he insists on this "minuteman" crap which always gives me a stomach ache. Learning to build our own solar panels and wind power equipment sounds like something our government would back because it would help us meet our goals to lessen our carbon footprint.

So you might want to watch this video to see if you really learn how to make a solar panel, etc. You will just have to tune out all the "Glenn Beck" garbage. I got sick of it after a while and so, since I don't really plan to build my own solar panel anytime soon, I cut the video short before I got to the actual instructions which you probably have to buy a book to get. If you have a stronger stomach than me, or you happen to be a "patriot" (as opposed to whatever the rest of us Americans are), then you can probably go all the way through and find out the steps you need to take to go "off the grid".

I forgot to save the link for the video I watched. Frank Bates does have a web site but I did not find that particular video on his web site. However, the web site does have the plans for building your own solar panels, etc. Just type the name Frank Bates in the search window to find the site.

I did find the link to the video called Power4Patriots:

http://power4patriots.com/video/

8.13.13

This Is the Droid You're Looking for!

Did you hear? That next new thing we're looking for to kick-start the future and, incidentally the economy, may have arrived. History may be about to repeat itself and this could be Henry Ford all over again, without the pollution. Elon Musk, inventor of the Tesla car, Pay Pal, and the Ironman suit stayed up all night (I'm guessing many nights) to design the Hyperloop, a people delivery system that works like a pneumatic tube and moves people at the speed of sound anywhere on earth that the tube system gets built. Talk about a small world. It would take 45 minutes to get from NYC to the city of your choice in California.

I have been looking forward to implementing "Jetson" style transport, but now that it seems possible (tiny hiccup - it will cost billions) that maglev travel could become a reality I am not at all sure that I would want to travel the earth the way my prescription makes its way from the drive-thru to the pharmacist. There will be glitches, which may or may not involve spectacular, or even oddly silent, deaths.

Apparently a team of people have already been at work building a system that is very similar to Musk's. Perhaps the document he has gifted the world with will help them tweak their design to make it more feasible. That there are a few bubbles of excitement in my gut either means that this might really be something, or that I am craving champagne. Oh, why not both?

9.5.13

<u>Going Bigger, Much Bigger</u>

These baby steps we are taking to stop climate change are not enough. Yes, I do believe in climate change. I will cast my lot with the scientists rather than the politicians and the oil men (the oligarchs). We need to go bigger. We need to go much bigger.

Solar panels

I have written about this before but I will keep urging that we act on some kind of Solar Panel Program for America. Perhaps it could be a program like the Energy Star Program which gave rebates if you updated to energy efficient windows, doors, and if you added insulation. There were also energy rebates for updated furnaces and appliances if they are more energy efficient.

Since solar panels are so expensive, the costs might be prohibitive even with a rebate. Perhaps each year there could be a lottery that would update a certain number of homes without up-front payment. Payments could be made by paying your old average energy fee and subtracting the costs of energy with solar panels and then applying the difference against the cost of the solar panels. I'm not an expert in economics or in designing either public or private programs, but I'm sure there is someone who can turn this idea into a successful program. In fact, there are programs that allow you to lease solar panels.

If all of our homes had either solar power, or if solar power was not viable, then perhaps a neighborhood could share and pay for a windmill or other wind capturer/converter; wouldn't that stop a lot of CO_2 emissions? I also saw that it is possible to make solar panels that also act as siding. We'd feel much better about our comfortable lifestyles (which we really would hate to lose) if we did not feel that in order to be comfortable we will end up destroying our planet.

Simplify, simplify

Americans like to go at life with passion, verve, and energy; and sometimes with ambition, chicanery, and greed. We are supposed to succeed, and in big ways, or we are seen as failures by ourselves and others. Everything in America is supposed to move upward and we have taken this to our hearts both figuratively and literally. We are climbers of mountains, riders of zip lines, bungee jumpers, and sky divers. We will work 80+ hours each week to climb the ladder of success, for corporate advancement, to be a leading entrepreneur, a sports or music figure, a doctor, artist, dancer, film star, inventor, lawyer, thinker, educator, or innovator.

But right now America is slowing down just a bit. Perhaps we are supposed to be slowing down, taking a breath, collecting ourselves, listening for the small voice of creativity that sometimes eludes us until we are in the shower, or driving home, or washing dishes. We have this big energy problem to solve. We use too large a share of the world's available energy.

While it is true that energy can neither be created nor destroyed, it can be changed into another form. The form that the energy we are using is changing into is CO_2 and that CO_2 is making our planet warmer. It is also true that the planet will not die off right away if it warms up. But the extra heat will have effects that we won't like. It already is. Perhaps you are not one of the people who feels bad about the extinction of species (like the polar bear), but some of us do. And whether we believe or not that the level of the oceans will rise doesn't matter because it is already happening. It does not really matter if you believe that climate change will bring more extreme weather because that more extreme weather is already here.

When I first heard that some states were breaking paved roads up and turning them into gravel roads I was shocked. Are they doing this because we are poor or because they are deep into their small government campaign? My guess is that both things are true. However, maybe for our own reasons we ought to think about simpler lives, less spiffy roads, really switching out our gasoline-powered cars for cars powered by electricity (once again there could be programs to help us switch).

We could just slow our lives down just a bit, not go at life like it's an obstacle to be overcome or beaten into submission. We could be more Zen, more jazz, and less MMA, less hard rock, or rap. We could still be modern with our jets and our high speed trains but our pace would just be less assertive, less aggressive. We'd still work hard, but maybe 40 hours a week will be enough or 60; not 80. If we slow down just a bit who knows how many things might occur to us. We might even get more exercise and stay fit. We could savor that coffee, smell those roses. And stop adding to global warming.

Whatever we do let's go big! Let's pick at least one strategy that will lower our fossil fuel use a lot and let's stay with it. I just do not have faith that my little blue bins are getting the job done. Keep in mind: we are not wealthy. We will need help if America decides to go big.

3.10.14

<u>Saving Earth - Tokamaks</u>

I have been waiting for a new clean energy source for some time now. I trust that the cosmos, the universe, the prime mover, God, or whatever power you choose to believe animates the world, would not let one energy source run out or become problematic without making a new one available. This view is not based on science for me; it is intuitive, totally an item of faith and it will be great for humanity if it proves to be true.

When I got **The New Yorker** magazine for March 3, 2014, I could barely contain my excitement. I do believe that newer, cleaner (although not necessarily proven) energy source is already being assembled in Provence, France. There are new acronyms involved and new vocabulary words we will have to learn: e.g. ITER for International Thermonuclear Experimental Reactor, Tokamak, Tokamak Seismic Isolation Pit, fusion (as opposed to fission).

"Eventually, physicists hope, commercial reactors modeled on ITER, will be built, too – with no carbon, virtually no pollution and scant radioactive waste. The reactor would run on no more than sea water and lithium. It would never melt down." (It seems to me that it could explode, but scientists don't seem worried about this.)

"Thirty-five countries," the author of this article entitled "*A Star in a Bottle*", Raffi Khatchadourian, tells us, "are invested in this project which is so complex it requires its own currency."

This project was supposed to be completed in 2010, but 2020 is now the projected date. It was proposed in 1985 by Ronald Reagan and Mikhail Gorbachev and now involves, besides Russia and the United States, the EU, China, Japan, South Korea, and India. Coordinating this giant project is both an exhausting nightmare and an exhilarating mental puzzle, apparently. The building to house the fusion reaction has to be built to exacting specifications and, although a vast array of electronic wires and conduits must be available once the machinery arrives there cannot be any holes or seams in the walls. The building is being constructed in an earthquake prone area – thus the need for the seismic supports. The machines that will go in the buildings to create the fusion reaction are being built in the various nations and must be built to exact specifications also. Then everything built in far flung places must fit together perfectly like a giant jigsaw puzzle. It's not easy to build a synthetic star. There is lots of room for error and any error could make fusion unlikely.

Giant electromagnets are needed to contain the reaction as no known material will. These must be manufactured on site. Money comes and goes as various governments or member nations change parties or leaders and therefore ideologies. Some governments back the science for a while and then withdraw making progress precarious and uneven.

If you want a clear explanation of the differences between fission and fusion see page 46 of the article in The New Yorker. It is a lengthy explanation, but quite clear,

Other smaller Tokamaks (as the structures for fusion are called) have been unable to produce enough energy to even equal the amount of energy put in to the reaction – when energy out equals energy in this is called breakeven energy. Of course we want a lot more than breakeven energy if fusion is ever to become a feasible new energy source. This building in France will be the largest Tokamak ever built. Will the reaction just miss the breakeven point? Will the energy produced surpass breakeven or will the experiment produce that "bottled star" that will just keep on giving us cheap, clean energy and allow us to say good-bye to coal and oil. Six more years seems like forever to wait and that may prove to be a very tentative completion goal.

There is a lot more of interest in this article. If you find all of this fascinating, see if you can locate a copy. It will read like science fiction I promise you. I think it will be magnificent if these scientists succeed.

6.11.14

<u>Great Minds United Might Save All</u>

Things have been slipping around here in America. Not that we are no longer living in considerable comfort, but we had been on a material "high" for quite a long time with only a few relatively short downturns until now. These days I am seeing small (and not quite so small) signs that we are letting some things go.

There are not quite as many food choices on some of our grocery store shelves. Is this a supply problem; are droughts and extreme weather making foods scarcer? We are seeing markets in middle class neighborhoods that no longer supply esoteric perishable imports for every foreign food as they had been doing until the recession. We have heard talk about high demand for exotic ingredients being unsustainable. We are being treated to a "buy local" campaign which obviously helps keep jobs in the area, but which is also placing dietary restraints on adventurous eaters. This campaign seems to be about more than saving local jobs.

We also are finding some grocery store chains putting pressure on people to eat healthier food, but given obesity rates and an expanded public health care presence this at least seems logical. Bread, for example, has been a real target of this campaign with whole wheat bread suddenly much easier to find on the shelves than the white bread we are all used to. While this is not slippage *per se*, having to give up foods we love is adding to our sense of having shifting ground on which to stand. We are also seeing prices for food climbing dramatically after being fairly stable for years, again reflecting shortages that can be attributed to those droughts and that bad weather over our farmlands.

We are seeing our roads getting worse by the year. Even when patched or resurfaced our road surfaces seem to break down faster than they once did suggesting that something is different. Perhaps in the same way that some are trying to make tasty croissants without butter (impossible) someone is trying to come up with a new tar formula that is petroleum-free. (I made this up because I think it might be true.) Perhaps the real reason why road repairs don't last is that same more extreme weather already cited, or just that we need more roads to meet increased traffic so the roads we have are overused.

The parts of our America that come under government control used to run like fairly well-oiled machines but it seems that services keep being dropped or cut back. Fire departments are being combined. Police Departments are also joining forces with the smaller towns being covered by the larger towns. Garbage collection is an exception. Our garbage workers do a great job of piling ever more garbage that will not degrade for centuries into landfills, for now.

Local governments no longer offer free tree services, even for trees they planted. Programs for seniors to help with weatherizing concerns like energy efficient windows, insulation and siding are gone. Even programs to build ramps for seniors are much pickier about who can qualify. Lead paint removal programs are only available when there are children in a household and even those programs are not available everywhere. And don't get me started on what is happening with our schools.

If we look to the wider world, strife seems to be just about everywhere, on every continent, except perhaps Western Europe, and, although China seems quite peaceful for now, we cannot say the same for all of Asia.

Our oceans are full of oily gunk and plastic, so much plastic that animals are losing limbs from getting wrapped in fishing line or that plastic baling tape or just the handles of plastic bags. Hard plastic is being ground into pellets that are the same size as sand. When animals like birds catch fish or other sea life to eat they scoop up plastic or ingest these pellets that are already inside of the sea food they eat. Their bodies can handle sand; they can't sort out all the plastic. They feed their babies food full of plastic. The parents die with their stomachs full of plastic and their babies die of starvation. These things that are dying are part of the human food chain. Extinctions are occurring faster and faster.

Our oceans are dying. We will not survive on continents surrounded by dead oceans.

Despite all that we see around us that seems to back up that feeling I have of slippage, we have this political party in America which denies it all, which doesn't want to spend money. And while I agree that we absolutely cannot afford to squander money, these folks have decided that we really need to slip backwards – back to nuclear families (that doesn't sound so bad, I guess, except how would they accomplish that goal), back before birth control, back before religious freedom belonged to anyone but Christians (Christians who can pass certain evangelical tests), back before labor unions, back before public schools, back before social benefits, back when you got health care only if you could afford it, back before government jobs, back before higher education became universally available, back before space exploration, back to only private medical research, back before the highway system, before long distance travel – and eventually we will find ourselves descending into the new Dark Ages. What will they do then with all that oil they are so hot to find? They apparently want to industrialize – but who will they sell their goods to?

I thought we would keep moving forward to a future in which we found a way to be kind to our planet and still be prosperous, although perhaps a bit less materialistic. While the Republican Way might get us there it will involve considerable pain and comeuppance for the middle class and that long descent into darkness – all quite undeserved as far as I can see. It seems we will be punished simply for not becoming wealthy, which was quite impossible as the wealthy were busy insuring they had a huge advantage in the matter of hoarding wealth.

The Daily Beast had an article on Wednesday, June 11, 2014 (link below) that suggested that Europe descended into the Dark Ages the last time the climate warmed – a phenomenon pretty much localized to Western Europe and having something to do with separating crops and sheep. How much greater will the dip into darkness be with the whole world in a warming phase? Millions died the last time, mostly from famine. How many billions will die this time? That's one way to save the planet, but I thought we could save it with our brains.

So you can decide if you want to let that constricting view of the future that the Republicans are offering us go into effect. I don't know if the Democrats can save us either. The two groups together, if they decided to lead and use their brains and stop the moaning nostalgia crap, might just be able to get it done, but I don't think we'll get to see that. What if all the brain power all around the world dedicated itself to solving our planetary dilemma?

What was that old saying – the last one left on the planet should turn off the lights. We may still be here on the planet when it, once again, goes Dark. I think I will start hoarding candles and matches.

Here's that link:
http://www.thedailybeast.com/articles/2014/06/11/when-the-weather-went-all-medieval-climate-change-famine-and-mass-death.html

6.17.14

Appealing to People in Ancient Lands

Persia lives large in my imagination. I picture it as a center of art and culture with beautiful buildings covered with colorful tiles and symbols, often floral patterns which we still echo today in our textiles, especially the paisleys and the bright colors. The carpets of Persia are still famous and we have remnants of these beauties in the carpets that still come from the regions around Turkey.

The markets are a part of that picture in my mind, with more patterned fabrics and carpets and the smells of exotic spices and foods, with fruits piled in wanton display and coffee sipped from delicate vessels. Persians loved beauty and surrounded themselves with it, at least as I imagine this ancient empire. They were poets and they were also mathematicians.

Syria and Iraq both encompass lands that once were part of Persia and share in this rich history. The Tigris and Euphrates rivers cross in Iraq and they are highways that also take us back in time to that fertile triangle our schools mentioned so often as the cradle of civilization, as a place where people could settle down and stop wandering as nomads because the fertile soil, refreshed periodically by flooding, provided a perfect spot for agriculture, and the resulting steady food supply provided the perfect circumstances for people to congregate in towns, which gradually grew into cities.

It is always sad when once proud and powerful people lose their nerve or their confidence and fade from the heights they once occupied. I remember experiencing this for the first time when I was quite young and watched the movie Lawrence of Arabia. I was shocked when Lawrence fell from grace and retreated for a while into a kind of depression. He was such a powerful man, however eccentric, and he had that larger than life thing going for him. It was obviously difficult for him to deal with setbacks and I will admit to shedding a few tears when I thought all was lost.

Today we have two nations, Syria and Iraq, once storied cultures within the Persian Empire; lands that will always have ancient roots. Iraq also contains those two great rivers mentioned above, the Tigris and the Euphrates, which meet at Baghdad and once were part of a Biblical land called Mesopotamia (meaning the land between two rivers). There is so much history here in these lands which have lately become battlefields. If this is what oil did, then oil has destroyed more than earth's climate.

It is so hard to watch these great nations caught up in petty sectarian and ethnic disputes. Instead of trying to restore the glory of their more halcyon days, or rather than even learning to interact so that their culture can stop self-destructing, just when we could use their help to unravel the modern dilemmas, these two nations have devolved into endless power struggles that are reducing these fine cultures to rubble. Once again there are tears.

I am certain that the various sects and ethnic groups have much more in common with each other now than they will have if they let the world in. Still, learning contemporary tolerance and their historical commonalities should be bringing these groups together to give their country power to negotiate with other modern nations and to live peacefully among them.

Authoritarian personalities arise in these two nations and may seem familiar from the past. There seems to be almost a cult of the strong leader in Arabic nations, an attraction to bossy leaders who are able to keep a lid on internal strife for a while, although usually at a cost. If these nations could find a way to convene and keep a representative government that tried to iron out differences and produce useful negotiated policies, these nations might find themselves free from despots for all time and ready for a new sort of Middle Eastern renaissance of those ancient cultures that were once so influential on the world stage and so beautiful to their citizens. Then they could surely find peace and a place among modern nations. (I know our representative government is not much of a role model right now with its divisive politics and its stalemate in our Congress, but hopefully we will become unstuck eventually, and our representative government has worked well for at least 250 years.)

We are in an environmental crisis, which was foreseen but which we ignored for far too long. This crisis could turn our planet from a natural paradise to a natural disaster. We need to clean up our oceans and we need to figure out how to wean ourselves away from oil and fossil fuels. You folks should have a say in these endeavors since any solutions will affect your economies and since you may have to find new sources from which to derive your economic health in the world of the future.

Please get a grip Syria and Iraq; we need your assistance. Stop warring within and bring that intelligence and creativity that gave us Persia and Mesopotamia to the table with those who are ready to try to save our planet.

2.16.15

<u>Here Comes the Future of Infrastructure</u>

It seems as if I am always imagining that some really innovative person/persons stumbles onto the "next new thing" that will put the future of American and of the entire planet on a firmer footing as we face the wages of our sins against our little home planet.

Sometimes I imagine that someone will find a new fuel source, or will make a scientific leap that literally launches us into the universe. Sometimes I look for signs that someone has learned how to do something that will be the bridge from the Industrial Age into a new Space Age or the Global Age.

I may imagine that it will be something elaborate like a new mathematical formula, Relativity 2.0, or that it will be the product of a singular genius now hatching in a womb near you or in a high school near me, or in a village on the other side of the planet (Einstein 2.0).

It could be a new attitude that turns humans into a species that always cooperates rather than one that worships its diversity (well that's not going to happen and it might be unbearable if it did). Maybe someone will find a biodegradable substance to replace plastic.

I think the two people in this article and video may have given us an answer that will allow us to make an enormous leap towards solving our energy problems; the ones that have us siphoning fuel from the bowels of the earth; the ones that have us causing geological havoc deep underground; the ones that have us pumping so many hydrocarbons into our atmosphere that polar bears are gasping as their habitat disappears, and baby sea otters have to be hand fed because our oceans are too warm to offer them their proper and natural diet.

These two people may also have found a way to jump start our economy big time; to solve our infrastructure and our jobs dilemmas in one fell swoop. I hope this proves to be a viable process and that we can get started on making this happen right away. I hope it does not end up being too expensive or that it will prove not to be feasible (to not work on a grand scale as it did on a smaller scale).

Here is the story and the video. See what you think:

http://www.globetoday.com/this-amazing-invention-will-change-the-world-forever/#

https://www.youtube.com/watch?v=qlTA3rnpgzU

4.13.11

<u>Some Green Please</u>

I love trees. In fact I have a huge old maple right outside my office window which I like to go on and on about. Of course, I have yet to see a leaf on my tree since I moved in, but I do have a chickadee who must be building a nest nearby and I have a pair of squirrels chasing each other through my leafless tree. I am trying to be Zen about this and just enjoy my tree day to day, but I can't help jumping out of my Zen and wishing for the leaves to hurry up and unfurl.

Picture our world without trees. I always picture that scene in **Rollerball** where the partiers take drugs and go out to torch the trees for entertainment. I remember when they cut down all the elm trees because of Dutch Elm disease. I remember when streets were like tunnels in summer with the trees making sun speckled beauty overhead. I can't help but notice how many of our streets are bare of trees. The absence of trees makes our neighborhoods appear colorless and leaves us with nothing to soften the hard concrete edges of our streets. We get wires on poles instead of trees. There is a tree I pass that has been holed right through the middle so the lines can pass through.

I hope we don't ever abandon trees. We have a perfect symbiotic relationship with trees. My eyes are longing to see every shade of green. So much for being Zen.

7.23.14

How Canada Geese Became Pests

I drove near a mini-gaggle of Canada geese today.
They live on a very small pond next to a very busy
highway. I was reminded of how rare and
wonderful it used to be to see Canada geese. I drove
miles to see them at the wildlife sanctuary. I turned
my eyes skyward whenever I heard that honking
sound and watched the V's streaming sometimes
north and sometimes south depending on the
season.

That was before we covered all the wetlands in the
area with parking lots, grocery stores, malls, banks
and phone stores. Now Canada geese are seen as
pests who live on any tiny pond left after the
wetlands were filled in. Car dealers like to put
ponds on their peripheries because ponds make
their businesses look less businesslike. Canada
geese collect near these small, but permanent bodies
of water. They collect in our parks if these parks are
near water.

It seemed lovely at first to share nature with these
large wild birds but then they started to have lots of
babies and to take over our public spaces and to
leave green piles of smelly waste everywhere.

It reminds me of when we turn farmland into suburban neighborhoods and then berate the farmers if their houses have a bit of peeling paint or their lawns are not manicured to perfection or if they keep odorous animals or if they try to hang out and schmooze in nearby Seven-Elevens.

We feel the same way now about those once rare and beautiful Canada geese. People buy cutouts of coyotes and stake them in the ground to chase the geese away. We are irritated when traffic has to stop on our busy streets in order to avoid making baby birds orphans or massacring them. I was horrified when I saw a goose family crossing a six lane highway. I covered my eyes. But people stayed put until they made it to the other side. People barely stop at red lights. How long will human patience stay on the side of the geese?

It is sad that we have lost our wetlands (and perhaps environmentally harmful) and it is even sadder that one of the few signs of this loss is the nuisance that Canada geese have become.

I still look skyward though to watch the skeins of honking geese as they fly over in the spring and fall. Will geese and man learn to accept each other or will we eventually lose these lovely creatures by deciding they are annoying us and thus dooming them to extinction? We don't have a good track record.

9.25.14

<u>Apparently Greed Makes You Stupid</u>

Why aren't we forming committees, think tanks, and community organizations to voluntarily curtail our energy-greedy lifestyles gradually rather than waiting for the big fail. Isn't that how democracy is supposed to work?

We could sit here and watch our climate change as we would watch the symbols come up in the windows of a slot machine – what will our overheated atmosphere do to us today? We could watch as the ocean water that already floods parking structures in Miami at high tide, floods these parking structure permanently. We could watch as our beaches are pushed inland like the beaches in Norfolk, Va. We can watch as the water cycle is disrupted, as torrential rains fall in some locales while other, once green, places dry and crack with drought. It'll be exciting. What does nature have in store for us next? And this won't just happen in America; it will happen all around the globe.

Shall we wake up one day in a freezing cold house because it becomes finally clear to all that it is our love of fossil fuels that is fueling our wild climate swings? Our cozy houses and our McMansions seem far less comfortable already when we lose power even for short periods of time; how much less homey will they seem when we lose cooling, heating, electricity, and hot running water for long periods of time, or permanently? Are you ready to give these things up for good? Do you want to be taken by surprise?

If we have no electricity we will soon have to do without even cold water. All the old springs used to have taps where you could fill containers with water in an emergency. All those old taps are gone. We truck in fresh water to people in modern catastrophes. What if that isn't an option? What if no water trucks appear anywhere in America (or even the world) some day? We have no individual access to water. Recently town water has been considered a human need and most people are now on the water grid. What if you suddenly needed a well? Do you have the slightest idea how to dig one yourself and how to line it to make the water usable?

How long will any trees on your land last if we run out of fuel? Do you have a wood-burning stove or fireplace? We have gotten rid of our primitive backup systems. They took up too much space, they didn't look nice, and we never used them anymore. Do we think it will be easy to get new old-fashioned systems up and running again? Those old heating systems burned wood. Burning wood will no longer be a good choice. Wind and solar will start to look pretty good.

Climate scientists suggest that our lifestyles will change drastically and sooner than we thought. Even if they are wrong about the timing we know in our guts we have to find ways to live comfortable lives without fossil fuels or face the new dark ages. Doesn't it make sense to discuss all this and make plans for a future that will not impact so negatively on the earth and therefore, eventually, on us?

We have a group of Americans who are incredibly fortunate and who have also used their power to make sure the blessings keep pouring down on them. When they say "make it rain" they are not doing a rain dance; they are passing laws, regulations, and rules that will cause money to rain down on them. But money won't help much when only a few people have it. And money will burn, but even billions won't burn for long. You can keep using your money to buy you a skewed government and you can use it to make the "rain" continue to line your pockets until the planet says "time's up". You can feel happy that we "the little people, the takers" are forced to live subsistence existences when you are not. Your edge won't last long either. If the grid goes down it takes us all down with it (after the fights over any remaining stockpiles end). What fun would it be even if you did get to lead a shiny lifestyle in a very dull world?

Although a climate change apocalypse would certainly make other problems pale by comparison, it is highly likely that countries with the most oil will be able to hold out the longest and that may furnish them with the advantages they would require to rule the world – a double whammy.

It doesn't have to be this way (and perhaps it won't anyway). Can we afford to take that chance? We know in our hearts that our planet cannot offer all 9 billion inhabitants (by 2050) the same level of comforts that some of us enjoy today.

Your wealth, used to reengineer our lives, rather than rig the system so you few can hold on to your old lives – would make all the difference. If your funds and your business sense and your passion could do this for earth we would not bother you again for a long time about how wealthy you are. Don't worry; you don't have to spend it all.

Environmental Diary

Part 5

Science Fiction and Environmentalism

And

Space

You're probably saying right now, "What does science fiction have to do with the environment and our environmental abuses. Why do apocalyptic movies and literature play a role? Space and our planet has no connection right now. We cannot look to space for answers. Well, we can, but they don't seem to be immediately forthcoming.

This is actually the inspiration part of the environmental diary. This is our hope and the foreshadowing of what may be humankind's great adventure. If not there is still much to be learned from books, movies, and even our limited excursions into space.

I was fortunate to come of age in a sort of golden age of science fiction writing. One book after another went the old-school version of viral and these are books we knew, even as we read them, would become classics. I don't think I crystallized my concern about the state of our planet until I read these books. I knew how polluted our local lake had become. I knew about smog over our big cities. I knew about acid rain and Love Canal and species extinctions.

The **Dune** books by Frank Herbert get my credit for really showing me how biomes intersect and how everything is interconnected, a pronouncement I cannot pronounce enough. The Freemen with their "still suits" repurposing their own bodily fluids to replenish scarce water resources on their desert planet, riding their sandworms through all the areas of their planet where only they dared to go, well they were my superheroes and they brought home to me the flagrant waste of resources in our real world.

All these books made me yearn for a space empire before there was ever a **Star Wars**, and some of these books even predate **Star Trek**.

Manifest destiny was the term created to describe our passion to explore every obscure corner of our earth, to grab all the land on the American continent between one ocean and the next. It's done. There is hardly a square inch of our world where man has not set foot. I believe that our drive for manifest destiny will eventually (if we can conquer the technology) propel us out into space as we try to get to know every square inch of the multiverse. I also believe space is an appropriate outlet for adventurers who need to take risks, who will wreak havoc if forced to be domesticated. We must find a way to go out into space.

If we do learn to travel and settle on other planets, the stories of our apocalyptic demise may subside. Going to space will not get rid of the necessity to learn how to be caretakers of our planet, but it may provide an off-world garbage dump so we can clean up the joint.

12.26.11

Reprise, Science Fiction Book Lists

*

7.20.12

Spacy Again

*

11.24.12

Is Apocalypse Inevitable?

*

1.8.13

Ground Control to Major Tom

*

1.25.13

Curiosity and Terraforming Mars

*

2.8.13

Racing the Apocalypse

*

2.15.13

It's Asteroid Day

*

12.26.11

<u>Reprise - Science Fiction Book Lists</u>

Have you earned your science fiction chops?

Everyone needs the classical underpinnings of a

good sci-fi reading background. It should include at

least the following, although some of these authors

have many more titles:

Robert Heinlein

· **Strangers in a Strange Land**
· **Starship Troopers**
· **The Moon is a Harsh Mistress**
· **Time Enough for Love**

Arthur C. Clarke

· **Childhood's End**
· **2001: A Space Odyssey**
· **The Fountains of Paradise**
· **The Light of Other Days**
· **3001: The Final Odyssey**

Ray Bradbury

- **Fahrenheit 451**
- **Something Wicked This Way Comes**
- **The Martian Chronicles**
- **The Illustrated Man**
- **The October Country**
- **Death is a Lonely Business**

Isaac Asimov

- **Foundation Books (Foundation, Foundation and Empire, Second Foundation)**
- **I Robot**
- **The End of Eternity**
- **The Robots of Dawn**

Kurt Vonnegut

- **Slaughterhouse-Five: A Novel**
- **Cat's Cradle: A Novel**
- **Mother Night: A Novel**
- **Breakfast of Champions: A Novel**
- **The Sirens of Titan: A Novel**

Douglas Adams

- **The Ultimate Hitchhiker's Guide to the Galaxy**
- **Dirk Gently's Holistic Detective Agency**
- **The Long Dark Tea Time of the Soul**
- **The Salmon of Doubt: Hitchhiking the Galaxy One Last Time**
- **The Restaurant at the End of the Universe**
- **Life, the Universe, and Everything**
- **Mostly Harmless**
- **So Long and Thanks for all the Fish**

Frank Herbert

- **Dune**
- **Dune Messiah**
- **The Dosadi Experiment**
- **Destination Void**
- **Children of Dune**
- **God Emperor of Dune**
- **Heretics of Dune**
- **Chapterhouse Dune**

Aldous Huxley

- **Brave New World**
- **The Doors of Perception: Heaven and Hell**
- **Eyeless in Gaza: A Novel**

George Orwell

- **1984**
- **Animal Farm** (not really sci-fi, more political allegory)

You could then branch into fantasy like Jordan's "**Wheel of Time**" books, Harry Potter sort of fits in here also. Perhaps the Carlos Castenades "**Don Juan**" books, the "**Star Wars**" books, and the "**Star Trek**" books might also make the list.

These books help define where our future will go. Do books predict the future or create the future? It doesn't matter. If you have a great sci-fi background you will be there. These books are mind-expanding, like algebra. OK, maybe you don't like algebra, but algebra is brain training. It sets up certain pathways in your brain that you might not develop from any other discipline (although geometry also helps, and physics). In terms of helping with logical thinking I don't think there is a more powerful tool than mathematics.

If you don't read these books you will also fall behind the cultural curve. Certain allusions will elude you. Who are the Aes Sedai? What is a space elevator? Who is Trillium? What are suspensor chairs? It will just go on and on, the number of potentially important things you will not know.

Here is the rest of my sci-fi "hits" list.

Carlos Castenades – becoming a "warrior"

· **The Teachings of Don Juan: A Yaqui Way of Knowledge**
· **A Separate Reality: Further Conversations with Don Juan**
· **Journey to Ixtlan: The Lessons of Don Juan**
· **Tales of Power**
· **Second Ring of Power**
· **The Eagle's Gift**
· **The Fire from Within**
· **The Power of Silence: Further Lessons of Don Juan**
· **The Art of Dreaming**

Kim Stanley Robinson

· **Red Mars**
· **Green Mars**
· **Blue Mars**

J. K. Rowling

· **Harry Potter and the Sorcerer's Stone**
· **The Chamber of Secrets**
· **The Prisoner of Azkaban**

- **The Goblet of Fire**
- **The Order of the Phoenix**
- **The Half Blood Prince**
- **The Deathly Hallows**

Margaret L'Engle

- **A Wrinkle in Time**
- **A Wind in the Door**
- **Many Waters**
- **A Swiftly Tilting Plane**

7.26.12

<u>Spacey Again</u>

I don't know why space fascinates me. It's scary. It's infinitely infinite. It makes me feel like an ant. It is also so beautiful it takes your breath away and it holds out possibilities that are equally breathtaking. It makes us think deep thoughts; how did we get here, are we alone. It makes us hope there is a God.

I don't want to go there. I do not have that gene that makes me crave edgy adventure. But I watch avidly whenever we send something or, with even more excitement, whenever we send someone, into space.

Recently physicists announced that they had finally isolated the Higgs boson. It was produced in Switzerland by colliding two protons (in the Hadron Collider). The Higgs particle decays very quickly which is why it has been, and continues to be, so elusive. To most of us the significance of this scientific discovery sounds sort of like; something, something, dark matter; blah, blah, gives matter its mass; yada, yada, the God Particle, or, as someone said, "the Godless particle" because, apparently it offers an explanation that shows a scientific basis for how the Big Bang could work without God.
http://m.startribune/opinion/?id=161956395
Jeffrey Weiss, "The Godless Particle

~~~~~~

~~~~~~~~~~~~

Scientists have long believed that there are multiple universes. Although my mind is totally boggled by the size of one universe; I am being asked to fit my brain around the knowledge that there is more than one universe. My mind is still digesting the existence of multiple galaxies. However, intellectually I comprehend what scientists are saying. Conceptually it makes my brain hurt to try to picture it. But anyway, here's what they had to say:

In the most recent study on pre-Big Bang science posted at arXiv.org, a team of researchers from the UK, Canada, and the US, Stephen M. Feeney, et al, have revealed that they have discovered four statistically unlikely circular patterns in the cosmic microwave background (CMB). The researchers think that these marks could be "bruises" that our universe has incurred from being bumped four times by other universes. If they turn out to be correct, it would be the first evidence that universes other than ours do exist.

The idea that there are many other universes out there is not new, as scientists have previously suggested that we live in a "multiverse" consisting of an infinite number of universes. The multiverse concept stems from the idea of eternal inflation, in which the inflationary period that our universe went through right after the Big Bang was just one of many inflationary periods that different parts of space were and are still undergoing. When one part of space undergoes one of these dramatic growth spurts, it balloons into its own universe with its own physical properties. As its name suggests, eternal inflation occurs an infinite number of times, creating an infinite number of universes, resulting in the multiverse.

These infinite universes are sometimes called bubble universes even though they are irregular-shaped, not round. The bubble universes can move around and occasionally collide with other bubble universes.

http://phys.org/news/2010-12-scientists-evidence-universes.html

~~~~~~

On August 5th a new Mars Rover called Curiosity will land on Mars, we hope. It will be observed by the Odyssey, an array in the sky over Mars. Odyssey was having some difficulties, but has apparently been repaired so we are still hopeful that we will be able to watch Curiosity land.

Curiosity will explore this unusual Mars mountain:

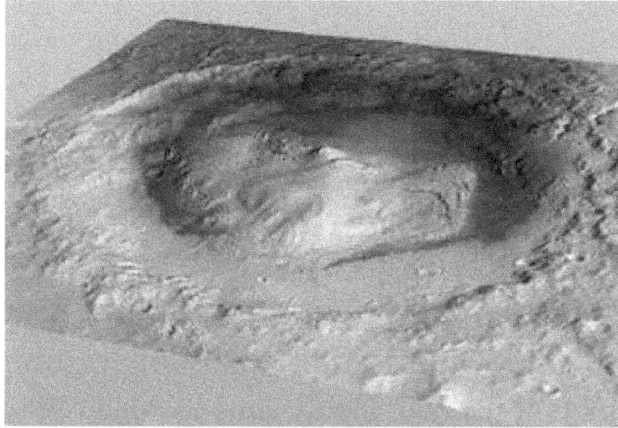

These photos are from the Christian Science Monitor

~~~~~~

~~~~~~~~~~~

We have to say good-bye to Sally Ride who left us this week. She was the first woman in space.

~~~~~~

11.24.12

<u>Is Apocalypse Inevitable?</u>

Most books and movies that paint a picture of the future of mankind give us a dark, empty, scary post-apocalyptic world that posits an argument that we are so bad for our planet and for each other that our only hope is to wait for us to destroy our planet and most of the people on it and start over with almost nothing. I had this conversation with my sister when we were leaving the movies recently and I must give her credit for crystallizing my own observations of the rather gloomy outlook presented by recent fiction. People obviously feel that we are on a path that will inevitably destroy us, and wound our planet, although not necessarily fatally. These stories hold out more hope for the regeneration of nature than for the survival of humans.

We had the **Road Warrior** series of movies with gangs of people in rags and stylish accessories creating strongholds to hoard fossil fuels and sallying forth to attack each other with vehicles that may have inspired robot wars; lethal vehicles driven by equally lethal skinheads who kill with little or no provocation and with an obvious sense of glee. They have one opponent, **Mad Max**, who is a vengeful hero who alone exhibits a sanity that seems to have deserted the rest of the bare planet that serves as the backdrop for these future wars. Only the children who Mad Max reluctantly champions hold any hope for a future less savage than the past or the present.

Cormac McCarthy's book **The Road** is perhaps the dreariest and emptiest of all the post-apocalyptic books. Is our planet killed by disease, war or pollution? We guess at but are not sure of which of the sins of man has produced this grim future in which a man, growing weaker and sicker, tries to escort his son to some unknown group of people who will offer him a chance to survive and start a new history for mankind. On the road they encounter small groups of survivors who will kill them if they can to help eke out a survival from the scant resources available. In spite of the inability to find any community of survivors who have retained any degree of compassion and unity these two trek on until they reach the Pacific Ocean and just as the father must leave his son, we get the tiniest hope that the human race and the man's son will go on.

Waterworld is certainly not any world that tempts us to create a similar future and **The Postman**, although it does inspire with the heroic attempts that are made to preserve a small aspect of the law and order the US government represented, also first shows us a world that has gone terribly wrong, a world where a thug rules by fear and savagery through a gang that accepts despotism as strength (where have we seen this model of human government before – answer – everywhere).

Even **Cloud Atlas**, that crazy quilt time travelling history of mankind, shows us losing the world as we know it through greed, self-indulgence, laziness and ennui to the amoral application of bad science to fuel our comfort. Starting over, by consensus, in each of these stories seems the only hope for our earth and for mankind.

It is no wonder we long to travel into space. A future that gives us the ability to populate other planets and form an Empire that takes in the whole universe seems to be the only future that promises any positive outcome where humankind with all of our flaws gets to survive for centuries and centuries more. Of course fiction is by definition not true, and this is us, putting our fears on screen and on paper. Is it a self-fulfilling prophecy or just a warning to change our ways? Will we heed the warning? Is the conquest of space the answer, not to changing human nature, but the answer to longevity for mankind? Right now that doesn't matter because we have not developed the ability to travel in space. What if we never do? What if this little planet is all we have? Can we avoid a future where we turn our dead into food (**Soylent Green**)? I assume we can, but will we? We have a lot of work to do to avoid the apocalypse and save our planet and ourselves. If we can't agree on short term policies how can we hope to invent a long term survival?

1.8.13

<u>"Ground Control to Major Tom"</u>

Of course you cannot put a You Tube video in a book, but if you want to listen to this excellent, now old-school, song just type this into your search window:

http://youtu.be/Wngk22chbq0

Romance your honey on February 14, 2013, and face annihilation from space on February 15, 2012. OK, I exaggerate. But another of those pesky asteroids is due to pass by Earth. This poor asteroid is so ordinary that it is only known as 2012DA14, no heroic Greek or Roman name for this space rock. It is about the size, say those in the know, of two train cars. The scariest thing about this asteroid is that it will pass between the Earth and the Moon and that just doesn't sound good. Scientists are not at all worried that they are wrong about the orbit, but they are worried that the asteroid may take out some of our satellites as it will cross their path of orbit twice. Even so, space is so, well, spacious that it is unlikely that any satellites will bite the dust.

Rocks that hurtle through space, or asteroids, as we have elevated them, are worrisome. Meteors and asteroids have hit Earth in the past, although no terribly dramatic examples have happened recently, thank you universe.

Here's what Gina Sunseri had to say on abc.go.com on this date (Jan. 8, 2013):

> *Weekly, sometimes daily, an asteroid zips close enough to the planet to show up on NASA's Near Earth Object List. The 99942 Apophis asteroid was once thought to be the one that threatened the Earth most, the one that could smash into the fragile planet. But scientists have had enough time to study Apophis to know it isn't a serious threat.*
>
> *Is the possibility of an asteroid hitting the Earth science fiction or science fact?*
>
> *Dr. Edward T. Lu might seem like an unlikely asteroid hunter. He's a physicist and former astronaut. For skeptics who think asteroid impacts are science fiction, he said, check what happened in Siberia in 1908.*
>
> *A 330-foot meteor exploded in the atmosphere above the Tunguska River with an impact 1,000 times*

more powerful than the atomic bomb that detonated on Hiroshima. The force in Siberia destroyed an area the size San Francisco.

Lu now heads the non- profit B612 Foundation, a group dedicated to hunting down asteroids before they hit Earth. B612 wants to launch the first privately funded deep-space mission: Sentinel, a space telescope to orbit the sun and map the inner Solar System in search of asteroids that smash into Earth.

The goal, Lu says, is to see what's out there. Before it hits Earth.

The problem, he said, isn't the asteroids that hunters know about, but those they don't know about. "For everyone we know about, there are about 100 more we don't know about," he said. "We have to find the other 99."

To do that, Sentinel will do what no government has funded yet, a dedicated long-term search with a unique infrared space telescope constantly scanning space for threats, asteroids large and small.

"Once we find an asteroid," Lu said, "it is possible for us to predict its trajectory. We know the government wants to discover asteroids big *enough to wipe out the planet but we also want to find those that could wipe out a city the size of New York, or Hong Kong, or Houston."*

The plan is to launch Sentinel in five years on a SpaceX Falcon 9 rocket, orbit for five years and gather information about asteroids that would give governments time to take action.

What kind of action? Scientists sometimes talk about the three Ds: detect, deflect and destroy. Lu Scoffs at the notion of blowing up an asteroid in space. It creates more space debris. Deflection is much more logical.

There's a bit more. Follow the link:

http://abcnews.go.com/Technology/asteroid-hunter-envisions-telescope-prevent-dangerous-earth-collisions/story?id=18157801

1.25.13

<u>Curiosity and Terraforming Mars</u>

We are human and we have a tendency to anthropomorphize everything. So, that little robot-science lab on wheels that is wandering around

Mars makes it almost seem as if we have a person on Mars. Curiosity is cute; it has one of the same eyes as Wall-E who had a delightful personality. By association we also suppose that Curiosity is equally delightful. Wall-E, of course, skewed male, while Curiosity has been presented, so far, as androgynous, making it difficult to choose an anthropomorphic pronoun. Whatever sex is assigned to Curiosity, I have decided to think of her as a "she" because I can. As I go through my day I sometimes remember that our little Curiosity is traveling ever so slowly over the landscape on MARS! She is constantly taking photos that she sends back to earth and she is sampling the materials on the surface of Mars to send us data about the chemistry of the Martian surface. And like Wall-E's girlfriend Eva, she is looking for any signs that there ever was or that there is now life on Mars.

Curiosity's newest photos show that water, good old H_2O once ran over the Martian landscape although she has not found any actual water. Photos show configurations that we know from long observation represent streambeds or trails carved by running water. There are rocks in the photos that show that the sedimentary action of water has acted on the Martian landscape at some time. This is all very encouraging to scientist who have long been discussing terraforming Mars.

Scientists believe that we could turn the thin Martian atmosphere thicker by purposely doing things we have done by accident which have thickened our earth atmosphere, or in other words by emitting greenhouse gases into the Martian atmosphere. Once a thick atmosphere was created we would use mirrors in orbit around the planet to melt the poles so they could release their water to fill the depressions on the Mars surface. We would then find a water cycle of evaporation and precipitation developing, after which pioneering life forms like mosses could be introduced and encouraged to cover the surface of Mars. You can see a demo of terraforming in this youtube video.

http://www.bing.com/videos/search?q=terraformin g+mars&view=detail&mid=D75373015FD04A7D544 4D75373015FD04A7D5444&first=0&qpvt=terraformi ng+mars

The ideas for terraforming Mars have been around for several decades now. Kim Stanley Robinson wrote a fictional Mars trilogy to help us picture the possibilities. Red Mars, Blue Mars and Green Mars, the three volumes in the trilogy create a set of Martian pioneer people who take us through the three stages of terraforming and the human drama that might accompany such an endeavor.

We may succeed in terraforming Mars or not but it always helps humans to strive for something beyond, something that stretches our minds and our talents and gives us a sense of some control over a massive and possibly indifferent universe.

2.8.13

Racing the Apocalypse

OK, well I have been thinking some more about our two alternative futures; the new age of technology and innovation vs. various apocalyptic scenarios. It seems to me that we are actually involved in racing against the apocalypse. We have become increasingly preoccupied with the end of the world.

I went to Wikipedia to find the number of movies that have had apocalyptic themes and the progression was clear and stunning. Before 1950 only 4 movies appeared on our movie screens with "end of the world" scenarios. By 2000-2009 the number increased to 55. In only one decade did the number of movies that portrayed world destruction decrease (in 80-89 there were 30 such movies, and in 90-99 there was one less, or 29).

of Apocalyptic Movies By Decade

(Information from Wikipedia)

Before 1950	**4**
1950-1959	**7**
1960-1969	**13**
1970-1979	**23**

1980-1989	30
1990-1999	29
2000-2009	55
2010-Present	17

Wikipedia also looks at all literature that depicts apocalyptic situations but that chart has too many entries and, because of the way it is organized, it is much more difficult to see trends. However there was another interesting chart which summarized the causes of apocalypse in fiction:

· Aliens
· Impact Event
· Disease
· Ecological
· Future Collapse
· Human Decline
· Monsters
· Sun
· Social collapse including economic and over-population
· Supernatural
· Technology
· War
· Unspecified

More than any other data this seemingly trivial stuff may indicate that we are at a turning point, that the world as we know it will experience either a thumbs up or thumbs down outcome in the near future.

We are racing the apocalypse, assuming it involves one of the catastrophic circumstances we can actually have some control over. We are still trying to prevent total global thermo-nuclear destruction (**War Games**). We are trying to understand the hostile forces arrayed against us in terms of "hater" nations or groups and to wend our way through a very tricky maze to an age of tolerance and peaceful coexistence. We are striving to end our dependence on fossil fuels so that we can avoid incinerating the earth and its inhabitants through global warming or drowning millions by melting the ice caps (or both). We hope our researchers are working everyday to wipe out disease and we are trying to take reasoned approaches to the exploding world population by saving species and working on a stable and safe food supply that will feed the 9 billion people who will populate the earth by 2050.

If we fail to innovate, if we fail at peace, if we fail to reverse the weather extremes that threaten our economies, our food supplies and our lives – if we fail in any of a scary number of ways – boom – apocalypse! No wonder we are so consumed with imagining all the ways we lose our way, the ways we lose a large percentage of the world population, and have to reboot the world with the survivors (and whatever they might randomly remember).

It all comes down to central heating for me. I love central heating (and cooling). I'm sure you figured this out about me by now. These are inventions that make our lives comfortable enough so that we can sit at our computers and write, so we can read and think and create, so that our lives become more than just a day-to-day struggle. So I hope someone saves us from the apocalypse so I don't have to live without my furnace or my solar panels. All right that's not a completely serious statement. I don't want to see the world go away or to see humans have to start over. I see human life as a continuum where we always improve upon the past and where we create solutions to any problems that arise and someday expand outward from the earth to the vast universe that surrounds us. I look forward to a new Age of Discovery. But I believe that right now we are racing the apocalypse.

2.15.13

<u>It's Asteroid Day</u>

Asteroid 2012 DA14 is due to pass by the earth today. The most interesting thing about this asteroid is how close it will come to hitting our little planet. It will pass between the earth and the satellites we have put in orbit around the earth. It will be only 17,150 miles above the earth at its closest point of approach. This morning we learned that while we are waiting for the asteroid and breathing a sigh at the near miss, unforeseen meteorites exploded over Siberia, stole 2012 DA14's thunder (literally) and injured at least 500 people with sonic booms that broke windows and shook up buildings.

We don't always remember that we are spinning around and hurtling through space in concert with many other space objects because it makes us dizzy and distracts us from our greatest gift, our lives, but we are occasionally reminded that we are tiny in the grand scheme of things and very, very vulnerable. I have no personal pictures of the asteroid that is paying us a visit today and I believe most of what we have seen on the internet are artistic renderings of our friendly neighborhood space rock. I find all of the things that happen in space intensely interesting, although I have no desire to go there and the mathematics that must be conquered by astronomers and astrophysicists is apparently beyond the capacity of my brain.

"An asteroid is coming! But don't panic. NASA says Asteroid 2012 DA 14 will make a record-close pass by Earth on February 15, but it won't hit us. Most asteroids are made of rocks, but some are metal. They orbit mostly between Jupiter and Mars in the main asteroid belt. Scientists estimate there are tens of thousands of asteroids and when they get close to our planet, they are called near-Earth objects."

"Asteroids have hit Earth many times. It's hard to get an exact count because erosion has wiped away much of the evidence. The mile-wide Meteor Crater in Arizona, seen above, was created by a small asteroid that hit about 50,000 years ago, NASA says. Other famous impact craters on Earth include Manicouagan in Quebec, Canada; Ries Crater in Germany, and Chicxulub on the Yucatan coast in Mexico."

"NASA scientists say the impact of an asteroid or comet several hundred million years ago created the Aorounga crater in the Sahara Desert of northern Chad. The crater has a diameter of about 10.5 miles (17 kilometers). This image was taken by the Space Shuttle Endeavour in 1994."

"What else is up there? Is anyone watching? NASA's Near-Earth Object Program is trying to track down all asteroids and comets that could threaten Earth. NASA says 9,672 near-Earth objects have been discovered as of February 5, 2013. Of these, 1.374 have been classified as Potentially Hazardous Asteroids, or objects that could one day threaten Earth."

"One of the top asteroid-tracking scientists is Don Yeomans at the Jet Propulsion Laboratory, which is managed by the California Institute of Technology. Yeomans says every day, "Earth is pummeled by more than 100 tons of material that spewed off asteroids and comets." Fortunately, most of the asteroid trash is tiny and it burns up when it hits the atmosphere, creating meteors, or shooting stars. Yeomans says it's very rare for big chunks of space to hit Earth's surface. Those chunks are called meteorites. (Ironically, that is exactly what happened in Siberia today.)"

This link will take you to the entire slide show on CNN:

http://www.cnn.com/2013/02/07/us/asteroid-approach-earth/index.html?hpt=hp_bn1

Environmental Diary

Post Script

Wrapping Up

A number of interesting and important environmental events are missing from this diary, probably because I was off in politics-land. One young man invented a revolving vacuum apparatus to get plastic out of the ocean. I saw it in my news feed on Facebook. It has already been installed and a two year experiment has begun. I have not seen any notices about the results achieved by this innovation but the two year time period is not up yet.

This diary never once mentioned the thick smog that has often choked Beijing, testament to the lack of environmental controls in China. Videos show people walking through the thick and probably smelly, toxic air with medical masks over their mouths and noses.

I followed the stories of the tiny but brave kayakers who went up against the giant oil drilling ships that were trying to leave the Seattle port for the pristine ANWR (Alaskan Wildlife Reserve) to do some off-shore drilling for guess what – oil. If you think oil is difficult to handle in warm seas, the Exxon Valdez oil spill already showed us that it is even more difficult in cold seas. The kayakers predictably failed to do more that frustrate the giant oil drilling ship, but fortunately for humans and animals who love cold climates, the oil companies found the conditions too harsh for their equipment. They vow to return.

Obama tried to enforce limits on CO_2 emissions by asking for companies who overshoot the limits to buy CO_2 offsets. If businesses find it too expensive to reuse the carbon dioxide they produce that is over the limit they can contribute cash to help others clean the extra emissions out of the air (although I doubt the monies are ever actually spent that way because the technology to clean air is not quite up to the job or the costs are prohibitive). Anyway, you can imagine the whining by the Republicans who don't believe climate science, about how these policies would kill more jobs. Carbon buy-backs or off set programs have not really gotten off the ground.

Most important of all, possibly, my diary does not mention the Paris Agreement (2016) reached at the Paris Climate Conference. This is an agreement with in the UN Framework Convention on Climate Change (UNFCCC) (oh, oh, the UN again). So far 195 countries have signed on to the agreement which "signals the turning point in the road to a low-carbon economy" say those who are optimistic. Many think this plan might succeed because each nation sets its own goals as it pledges to reduce greenhouse gas emissions and keep making the goals bigger. Each nation further agrees to try to keep warming below 2 degrees Celsius.

The agreement has aspects that are binding and some that are not. This development is worth following to see what the various signatory nations accomplish. In fact this agreement went into effect today, November 3, 2016, as I write this, according to PBS News. Experts agree that this agreement, while universally accepted, has no teeth and is nowhere near big enough to get us to where we need to be. The explanation given was that we would have to remove 15 times all the cars in Europe from the roads to keep global warming below 2 degrees Celsius. How likely is that to happen unless we invent a radical new way to transport ourselves that produces no carbon emissions and everyone uses it? Where is a transporter room when we need it? Everywhere I go I am surrounded by Chevy Silverados.

This environmental diary is done but none of our problems have been resolved. There have been good attempts to switch energy sources and clean up our messes. They are not big enough. The scale of our repair activities is too small and too haphazard and there is so much opposition. Even if we all pull together we may not undo all the damage. We may have to make accommodations that will allow us to live with climate challenges we have never faced before. And there are plenty of people who will only get very angry if they happen to even begin to read this book. Perhaps we need to work on being open-minded just as much as we need to work on innovation.

N. L Brisson writes for her own website **The Armchair Observer**. She has a BA degree from SUNY Potsdam and a Master of Education from the University of Arizona at Tucson. She cannot help her liberal leanings; they were an end product of her entire life. She lives in central New York, in a place with very snowy winters.

Please visit my website at http://thearmchairobserver.com/ where I have written about these and other subjects many times.

Or visit my Facebook page at https://www.facebook.com/The-Armchair-Observer-1113923478650446/

Photo Credits:

Cover Photo purchased from Shutterstock

Title Page Photo purchased from Shutterstock in color, printed in B & W

Dedication Page has a family photo

Cheap seats (usually blue) purchased from Shutterstock

Trees purchased from Shutterstock

All other photos come from Google Image Searches except p. 48 and p. 257 which are my own photos